Reviews for *That Book* series and David Conley

A totally delightful, energetic and imaginative book. For lovers of the ancient world, young and old!

Ursula Dubosarsky, Author of over 50 books and Australian Children's Laureate 2020-2021

Whenever Terry and I take our time-travelling rubbish bin for a spin through the ancient mythical lands of Greece or Midgard, we always take David Conley as our guide- and if he's not available then we take one of his books. He's the most knowledgeable expert on Greek and Norse mythological figures we've ever met, and we've met them all!

Andy Griffiths, author of over 30 books including the critically acclaimed Treehouse series

Almost as big and bright as me.

The Sun, centre of the Solar System

Some of the best books I have read and very funny.

Aaron, 8

David Conley is a bigger legend than Odin and a bigger marvel than Marvel! He puts the Rock into Ragnarök and the Norse into Norsænivegavinnuverkfæraskúrinngullbringur!!!! This is how the Greeks would have illustrated their myths and legends and it's what the Norse gods really looked like! Trust me, I've asked them.

I loved reading his mythologies - though I had to build a wooden horse and hide inside it to sneak into my son's room to get the books off him to read them.

Dr Craig Cormick OAM, award-winning author of over 40 books for the young and young at heart.

Doesn't say enough about my rings in his books. It's ridiculous. I've got heaps of rings and they barely get a mention.

Saturn, second biggest planet in the Solar System

I like all the short stories... I really like it... I never want to stop reading it... I just want to read it forever... it's really good.

Harry, 5

Definitely my favourite book that I've read then read again... Mum when is the next book out?

Will, 7

Your books are the best! I read them over and over, Mum says they are value for money. Your drawings have so much detail and the characters seem so alive (even though some are actually dead). I love learning about mythology and I can even pronounce all the titans and gods names better than my Dad (and spell them correctly too).

Josh, 8

I don't really look like that do I? I always thought me and my buddies were cute and cuddly and not at all terrifying. Could be wrong, though.

Black Hole, most powerful force of gravity in the universe.

That David Conley is a fantastic writer. All families should have his books. They're perfect for anyone aged 5 to 2 761 years, or when they die, whichever comes first.

Giselle, 12

As a teacher librarian, I am always looking for ways to engage students in reading and even reading about non-fiction topics. Ever since I have had David's books in the library, well they are never really in the library because they are always loaned out. What can I say? Children are really drawn to the imaginative ways these books are presented.

Marion, Teacher Librarian

There's no better feedback for this book then seeing a contented 7 year old boy curled up on the couch engrossed in the hilarious illustrations and intriguing tales. In his words the book is 'very unusual and very interesting'. In my words it's a unique, colourful and irreverent spin on some very old stories. He turns dusty into dazzling. I don't know many Greek myths, and I blame my misfortune on being born before David was sharing his talent with the world!

Emma, Playwright

I am way bigger in real life.

The Universe, the place where everything we know has ever happened actually happened.

First published 2023.

Dedicated to my three favourite people.

First published 2023

Dedicated to my three favourite people.

That Book About Space Stuff

David Conley

Let’s take a look around the universe...

Solar System Stuff

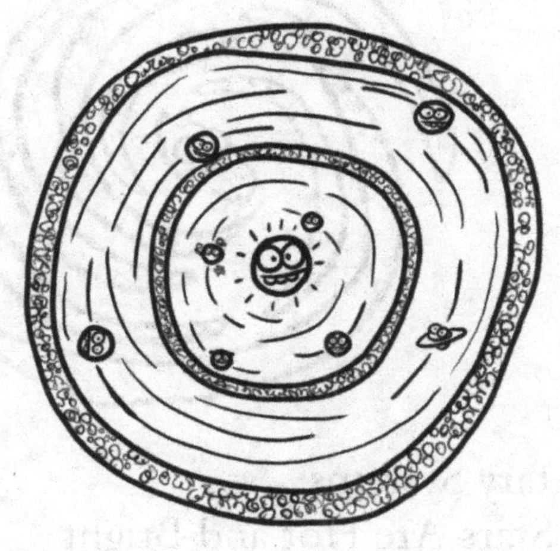

Galaxy Stuff

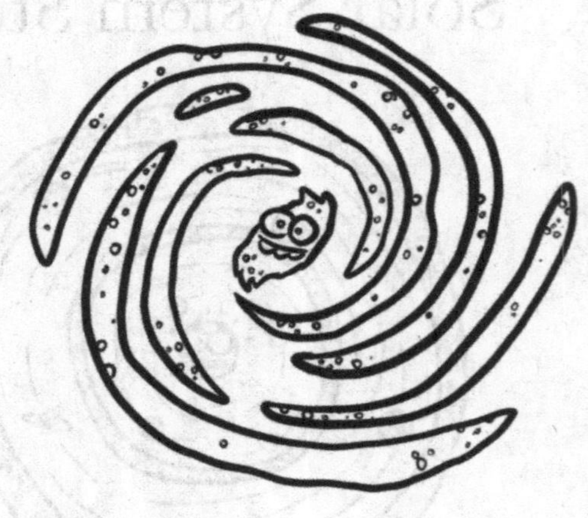

Universe Stuff

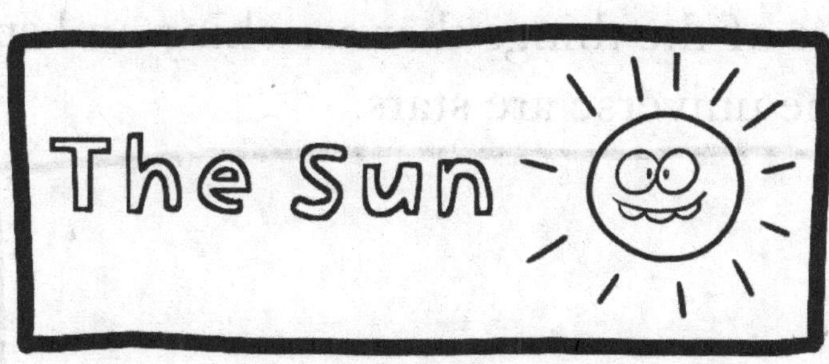

The Sun is a star.

Most of the things that are shiny and sparkly in the universe are stars.

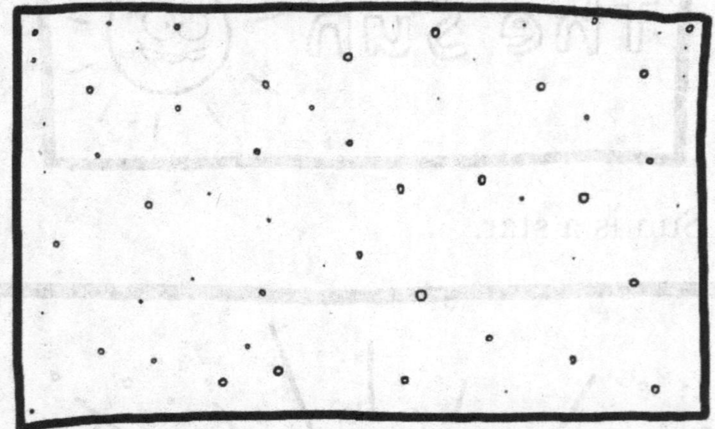

And lots of them look like the Sun up close.

Most stars are round.

Bright...

The Sun

Really, really big...

They're made of this really hot swirly stuff called plasma...

And they also have lots and lots of mass.

Things with lots of mass would be really heavy if we weighed them.

Because stars have lots of mass they have really strong gravity.

Gravity is a force that pulls stuff in.

Things with different mass to each other also have different strengths of gravity.

Things with weak gravity and strong gravity get near each other all the time...

When this happens, they both pull on each other. It's like tug-of-war in space.

But whatever makes the strongest gravity wins and pulls the other thing towards it.

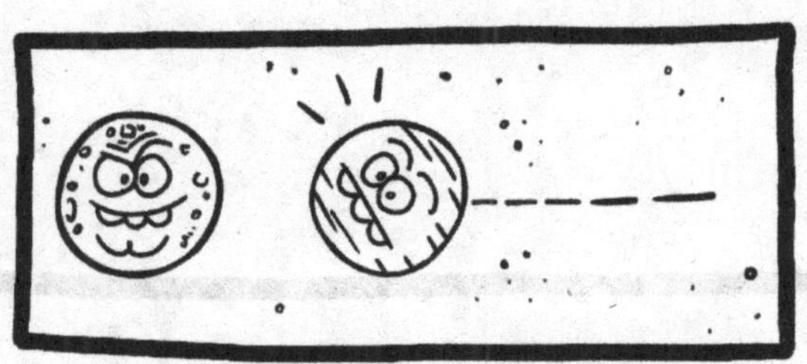

Stars can be pulled in by each other's gravity and make a big group called a star cluster.

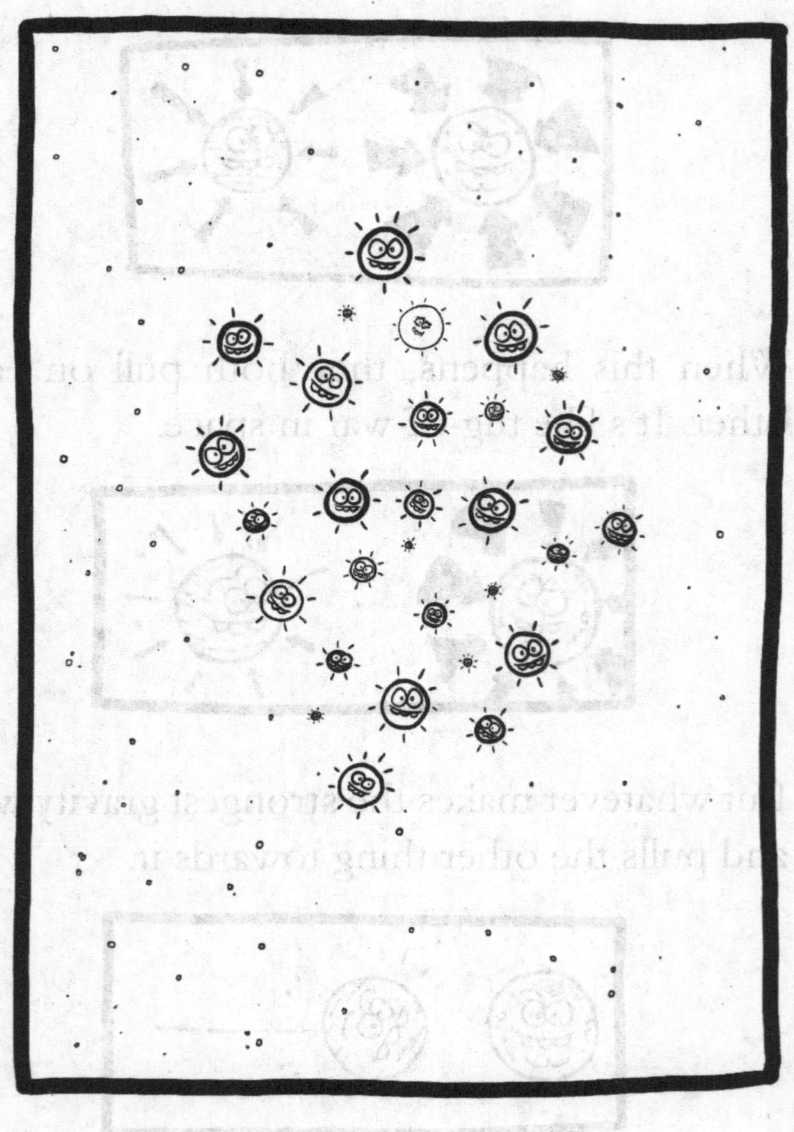

Like all stars, the Sun has really strong gravity.

And it pulls lots of other stuff towards it.

And that stuff goes around and around the Sun on a path through space...

The Sun and everything that goes around it are all a part of the Solar System.

All planets go around stars (when stuff goes around other stuff in space we call that orbiting)

All planets are round like balls.

And they go on an orbit around a star with nothing in their way.

If something gets in the way of a planet...

The planet's gravity will first pull it in...

Then it will either make it crash, force it to orbit the planet…

Or the planet's gravity will suck it in and shoot it out again like a slingshot.

The Solar System has eight planets. They are Mercury, Venus, Earth, Mars, Jupiter, Saturn, Uranus and Neptune.

The Planets

All the planets move on different orbits around the Sun and they're all made of different stuff like ice, rock and gas.

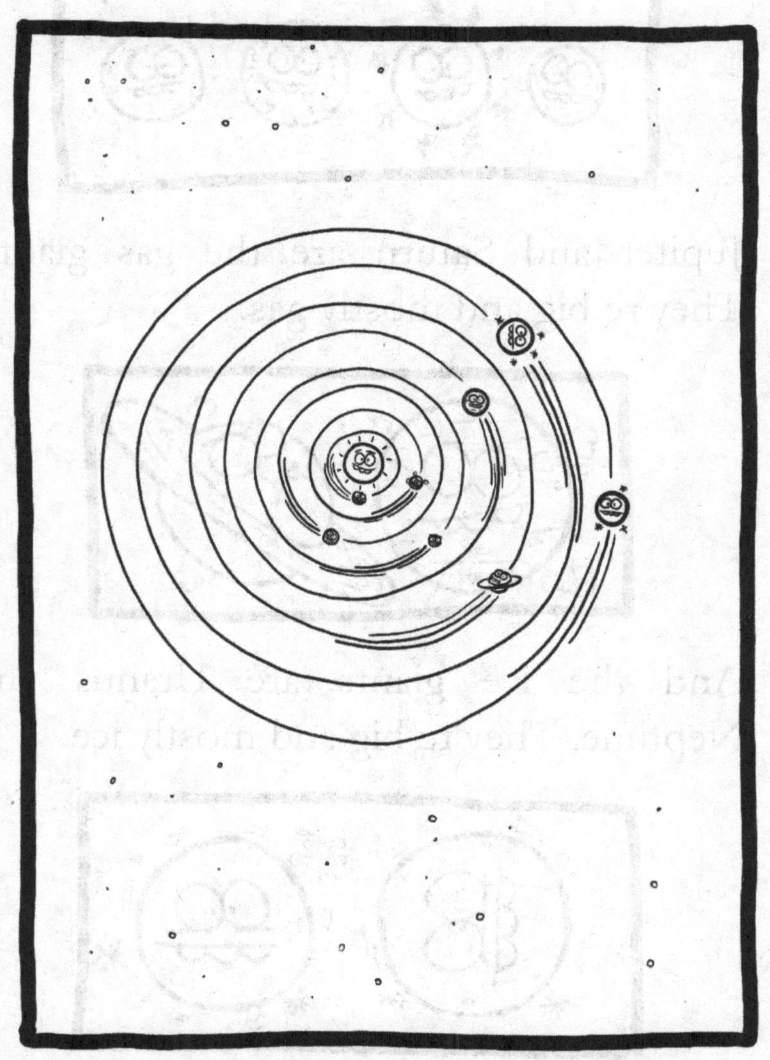

The rocky planets are Mercury, Venus, Earth and Mars. They are pretty small.

Jupiter and Saturn are the gas giants. They're big and mostly gas.

And the ice giants are Uranus and Neptune. They're big and mostly ice.

The Planets

Mercury is the closest planet to the Sun and Neptune is the furthest away...

Venus is the hottest planet, Neptune is the coldest...

Mercury is the smallest planet and Jupiter is the biggest.

Moons do not orbit stars.

They orbit other stuff like planets, dwarf planets and asteroids.

Moons

Moons can come from dust and gas clouds around something...

That pull together in a moon-ish shape.

They can also be free moving stuff that has been pulled into orbit around something.

Mercury doesn't have any moons.

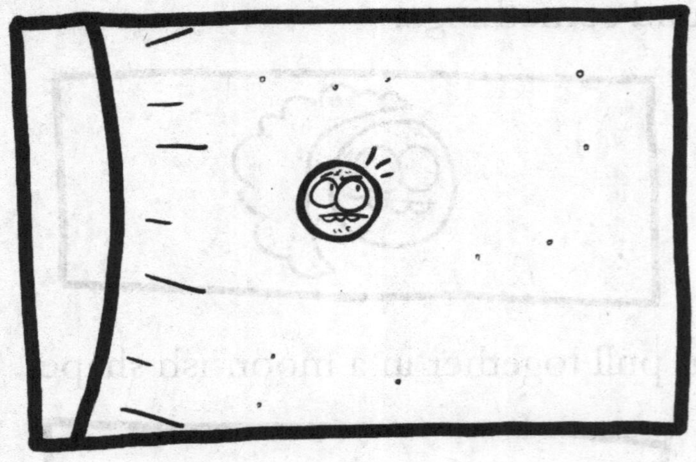

Neither does Venus.

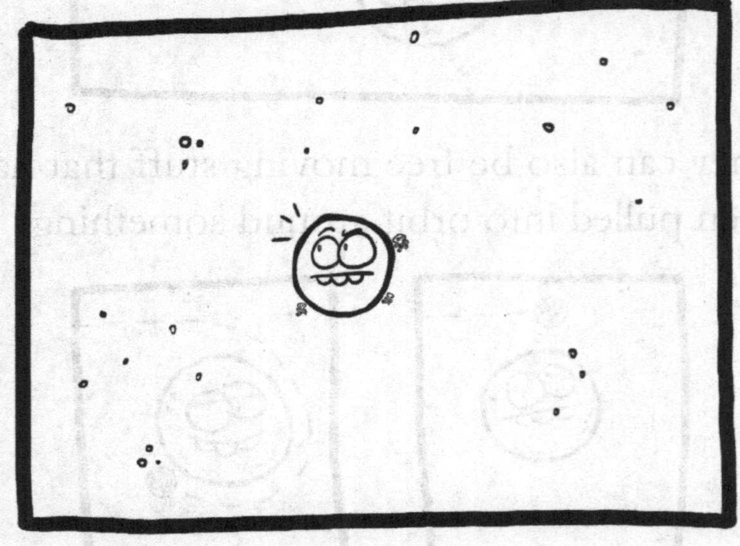

Moons

Earth has one moon.

Mars has two.

So far we know that Jupiter has seventy nine...

Moons

Saturn has eighty two....

Uranus has twenty seven....

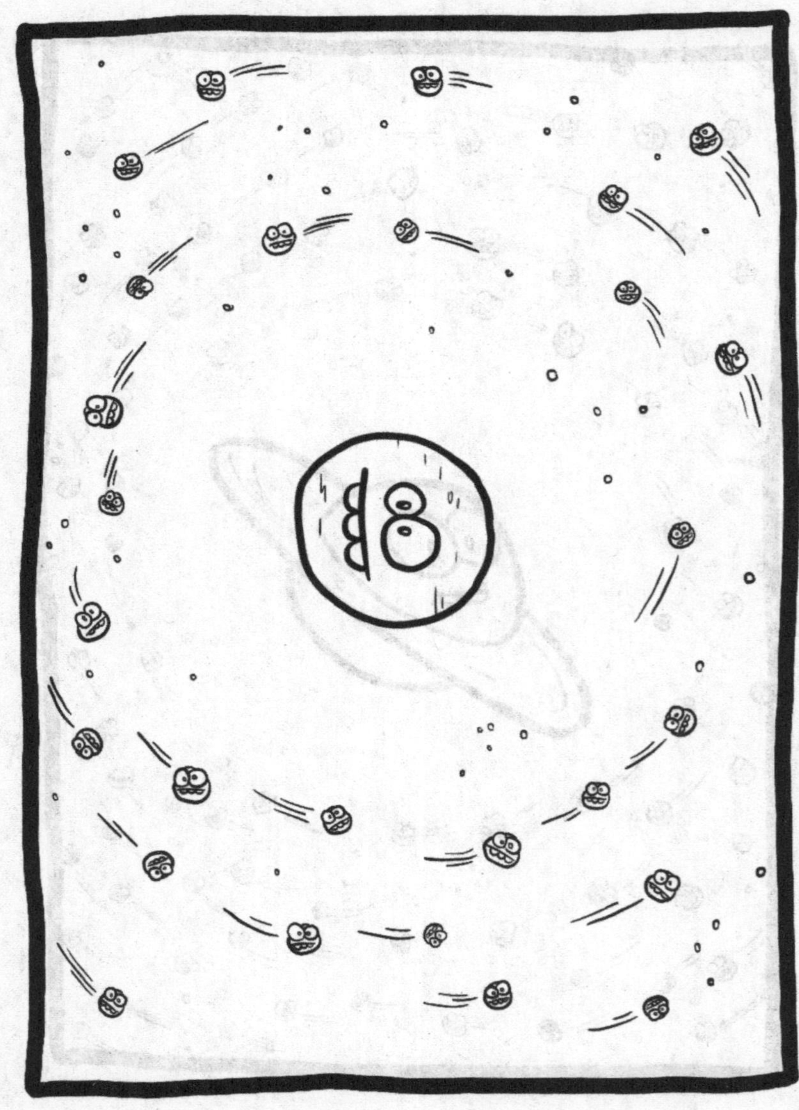

And Neptune has fourteen.

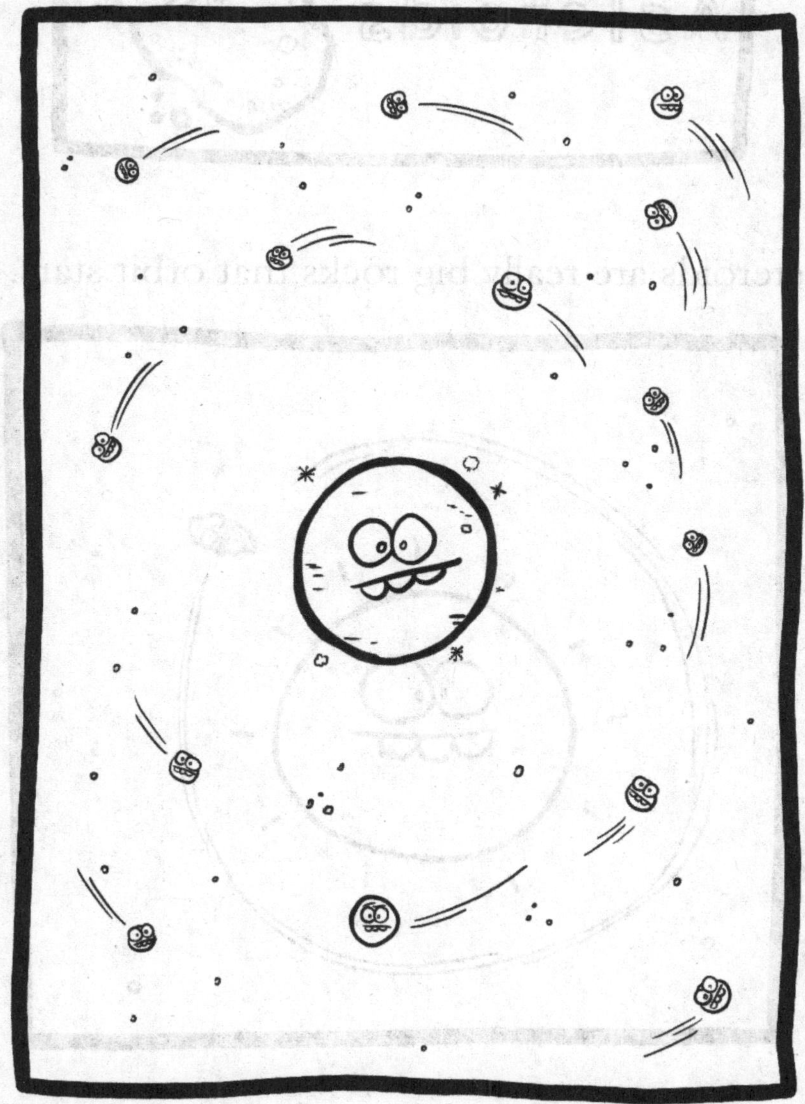

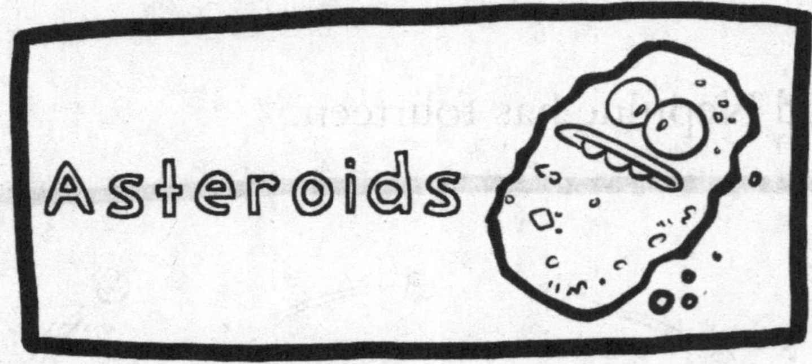

Asteroids are really big rocks that orbit stars.

Asteroids

They're not round like planets.

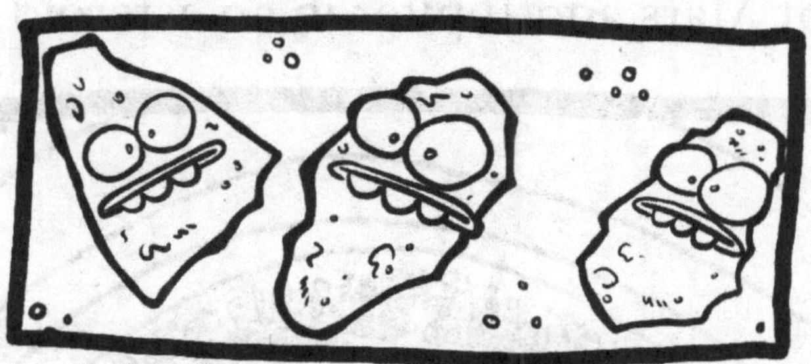

They're a lot smaller...

And there are lots of them in the Solar System.

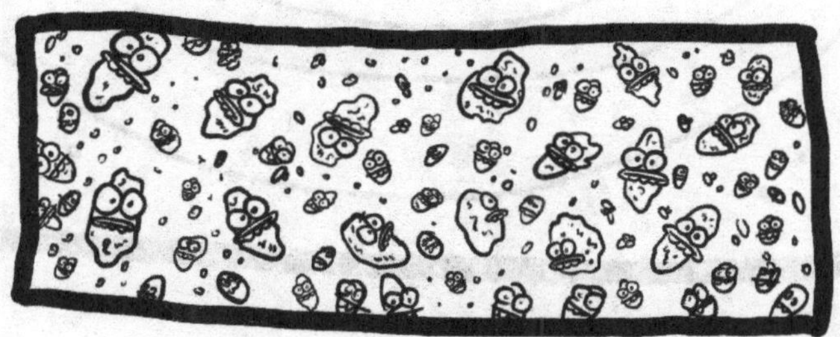

Most of the asteroids go around between the orbit of Mars and Jupiter in an asteroid belt.

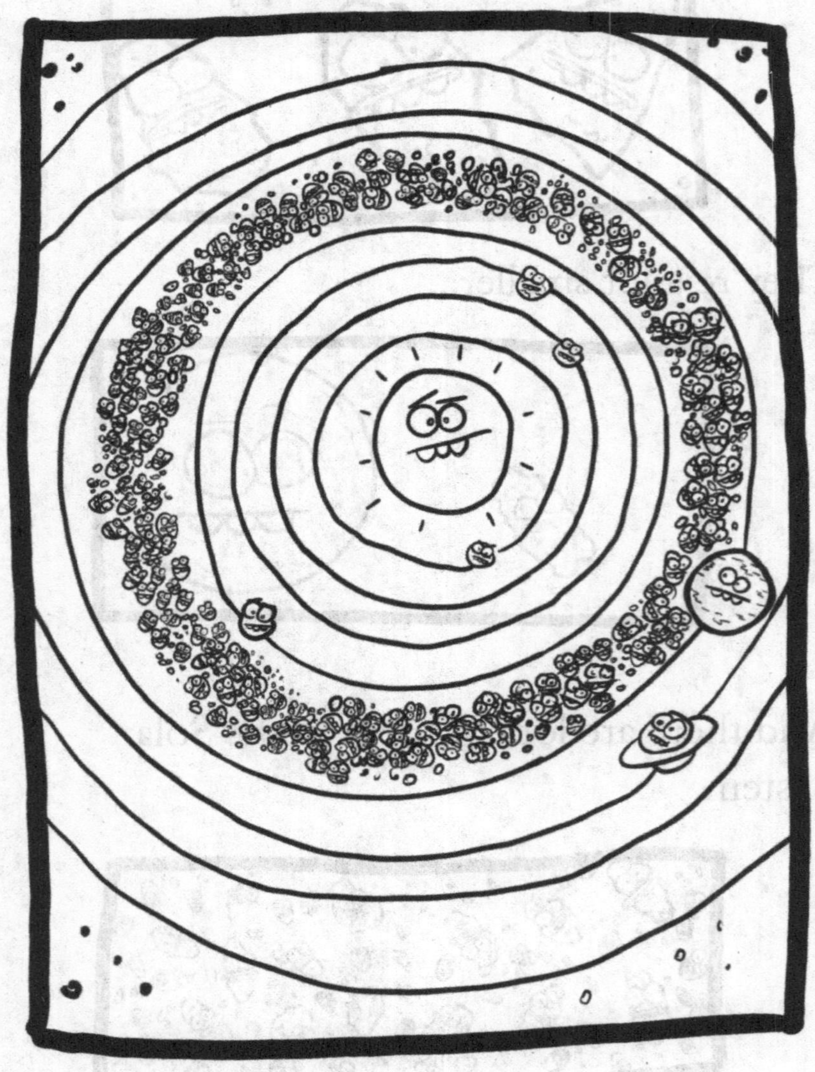

There's also a group that goes around in front of Jupiter on its orbit.

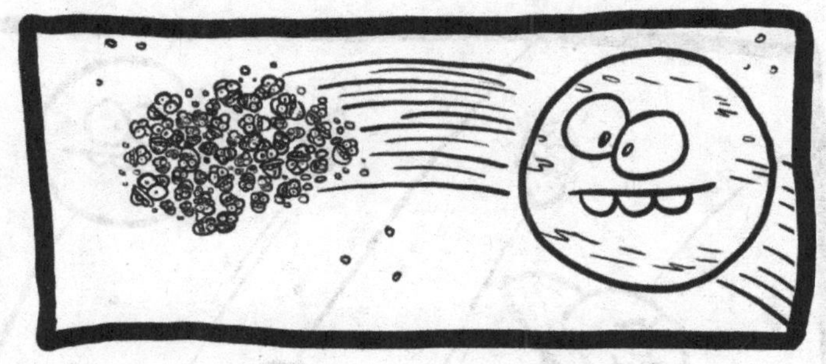

And a group that stays behind Jupiter.

Centaur asteroids are different to other asteroids. They're mostly made of ice, not rock.

And their orbits go through the orbits of Jupiter, Saturn, Uranus and Neptune.

Sometimes an asteroid gets too close to a planet...

Asteroids

Get pulled in by the planet's gravity...

And is stuck orbiting the planet for good.

Comets look a lot like asteroids.

But they're made of ice instead of rock.

Comets

Comets orbit stars.

When they are passing really close to the star they are orbiting...

The star's heat melts bits off the comet.

And all the bits melting off it make it look like it has a big, glowing tail.

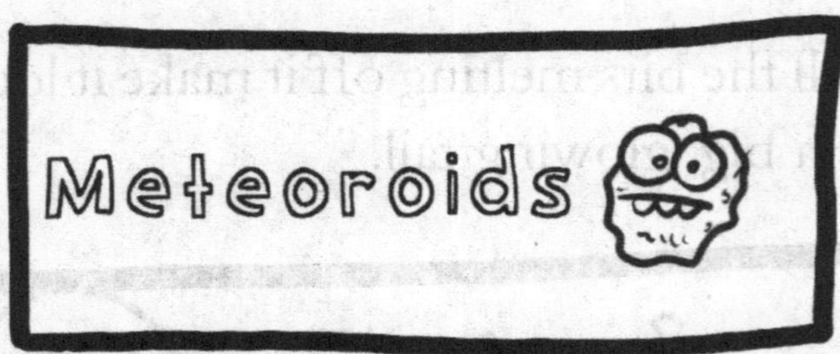

Sometimes asteroids run into each other and big chunks fall off them...

And sometimes chunks come off comets as they melt in the Sun's heat.

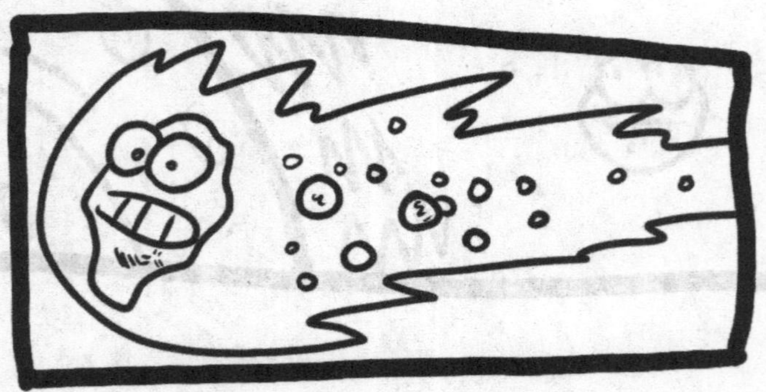

These chunks are called meteoroids.

Meteoroids orbit stars just like comets and asteroids.

But they are much smaller than either of them.

Lots of meteoroids get too close to Earth and get sucked in by its gravity.

When they enter the air around Earth, they become meteors.

Most meteors burn up in the air around Earth and are gone.

But some make it through the air...

And hit the surface of Earth...

The meteors that do make it to the surface become meteorites.

This is a micrometeoroid.

Micrometeoroids are much, much smaller than meteoroids.

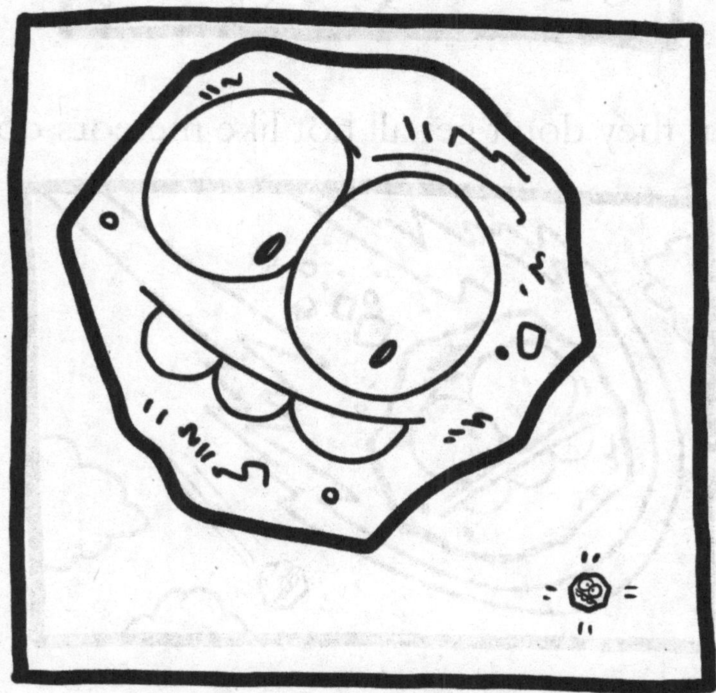

Lots of them enter the air around Earth and become micrometeors.

But they don't get all hot like meteors do.

That's because micrometeors are too small for Earth's air to heat them up.

So instead of getting too hot and falling apart in Earth's air...

They usually make it to the surface in one piece and become micrometeorites.

All the planets in the Solar System go on their own orbits around the Sun.

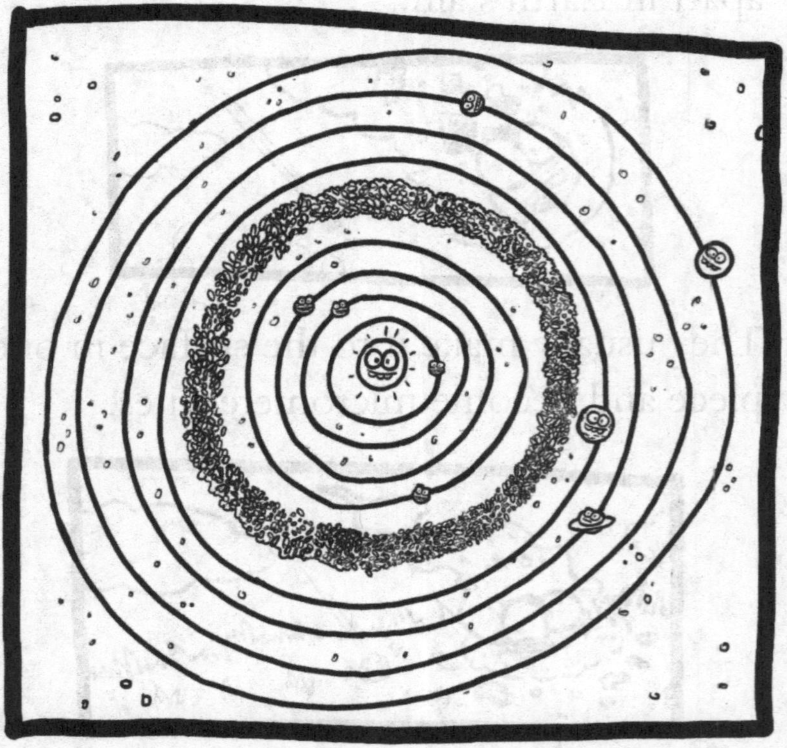

Kuiper Belt

Outside of all those orbits is the Kuiper Belt, a really big ring-shaped space...

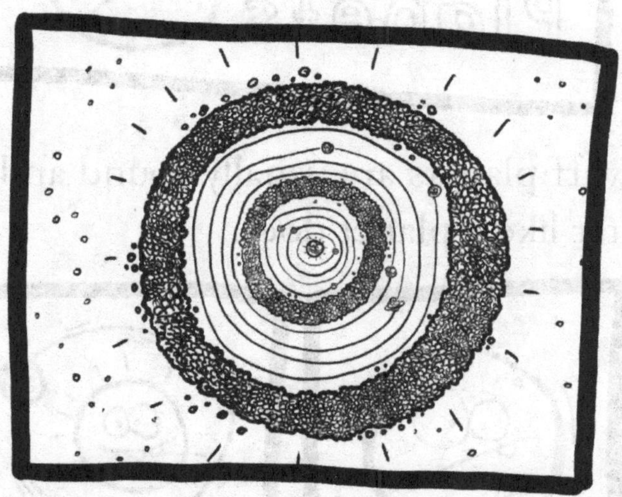

Which is mostly made of big ice chunks.

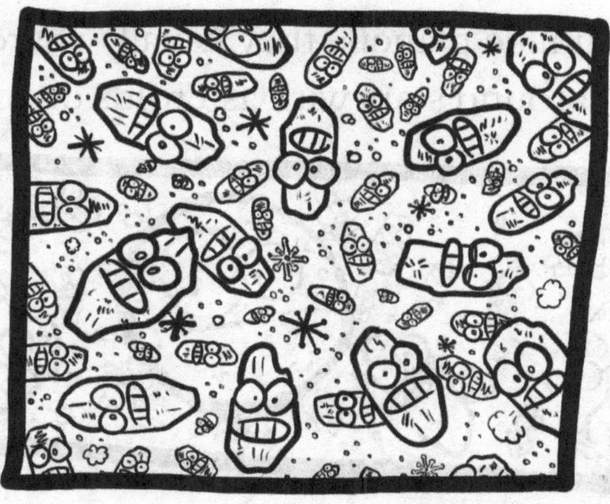

Dwarf planets are usually round and orbit a star like a planet does.

But, unlike planets, they usually share their orbit around stars with asteroids or comets.

Dwarf Planets

The Solar System has five dwarf planets that we know about so far.

Ceres is the only dwarf planet that goes around in the asteroid belt.

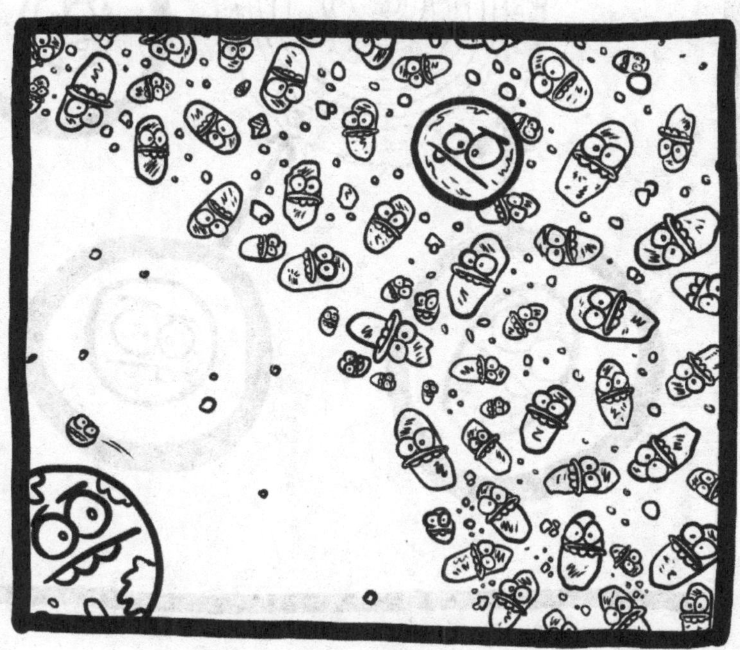

The other four (Makemake, Pluto, Eres and Haumea) are out in the Kuiper Belt.

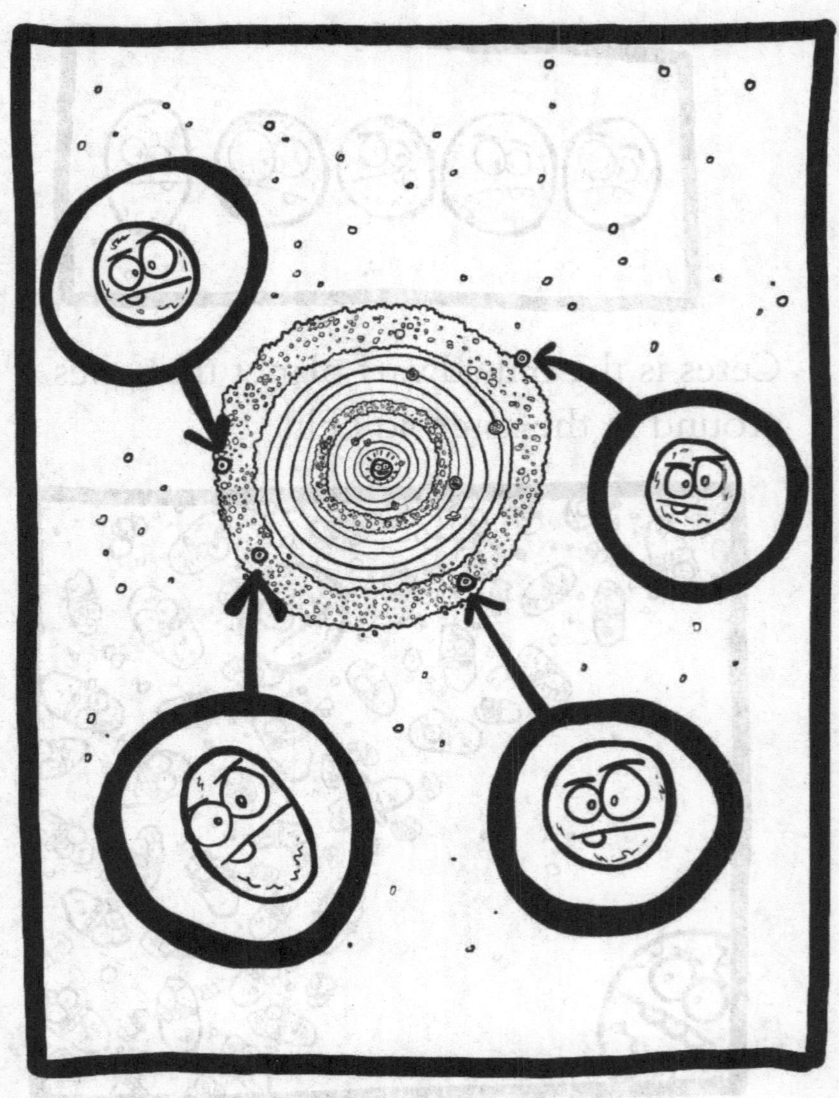

The Kuiper Belt sits around all the orbits of all the planets and the asteroid belt...

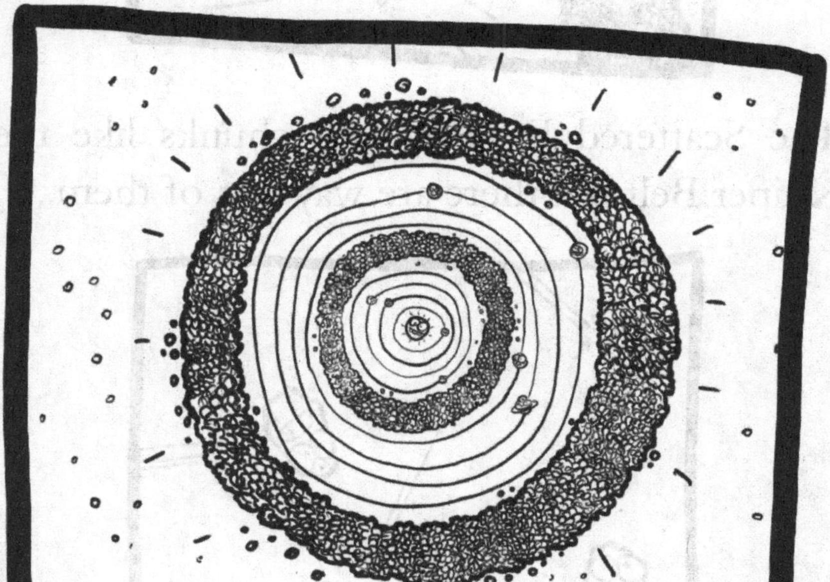

And if you go out past that you reach this space called the Scattered Disc...

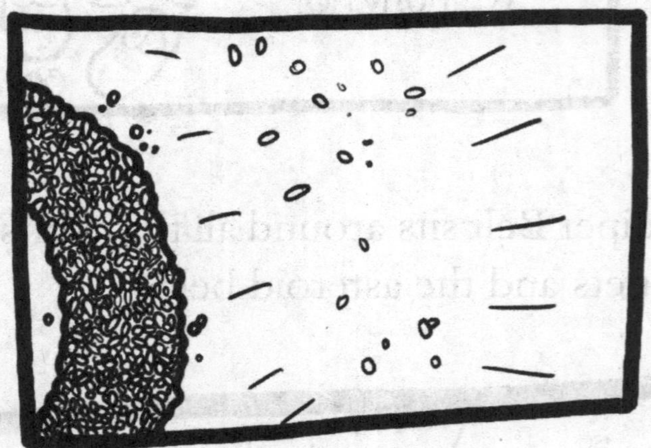

The Scattered Disc has ice chunks like the Kuiper Belt but there are way less of them...

The Oort Cloud

And past the Scattered Disc there is another layer of ice chunks...

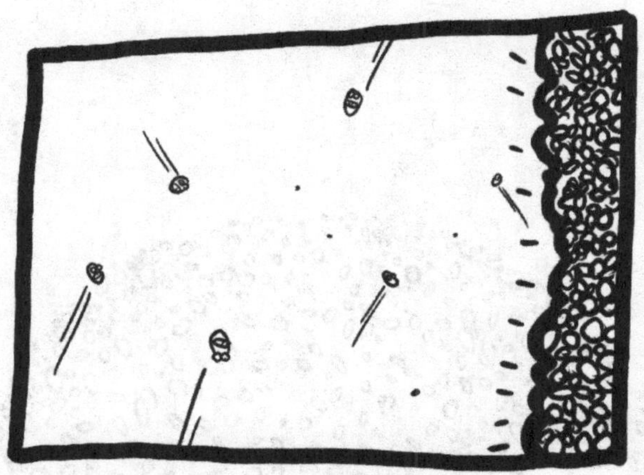

We're pretty sure this layer goes all the way around the Solar System like a big bubble...

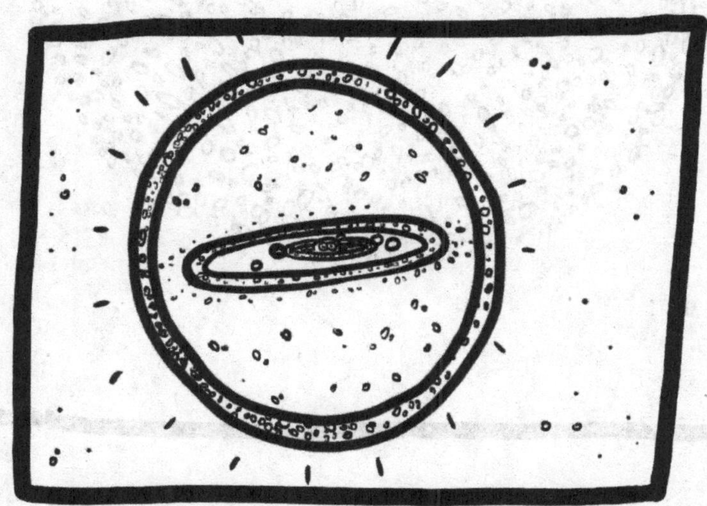

It's called the Oort Cloud.

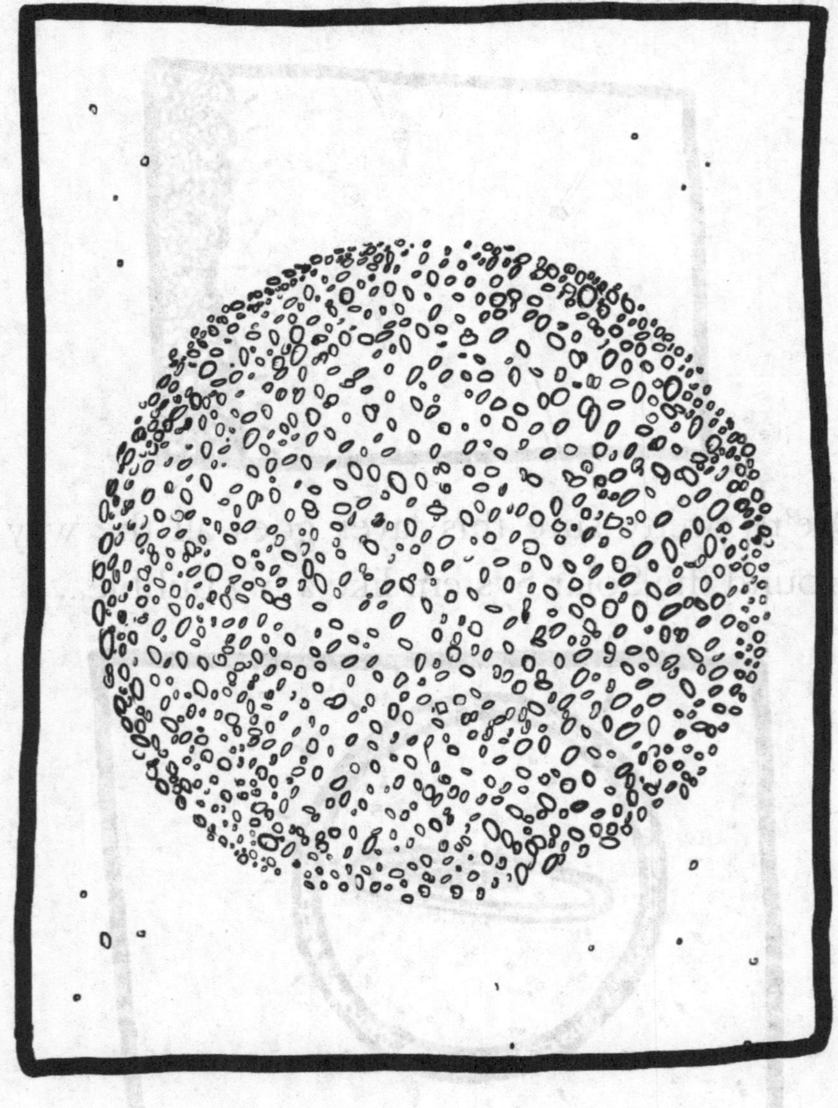

Outside of the Solar System there are lots and lots of other stars.

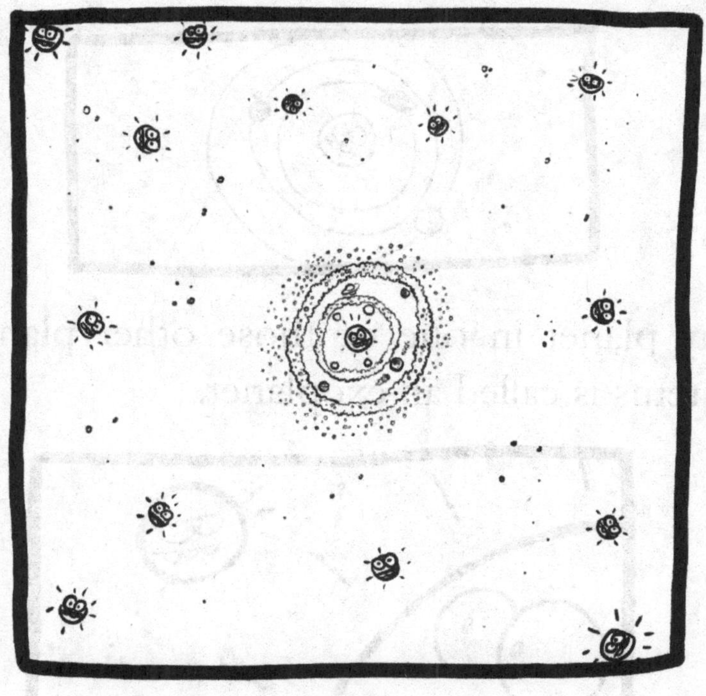

Plenty of those stars have planets and other stuff orbiting them.

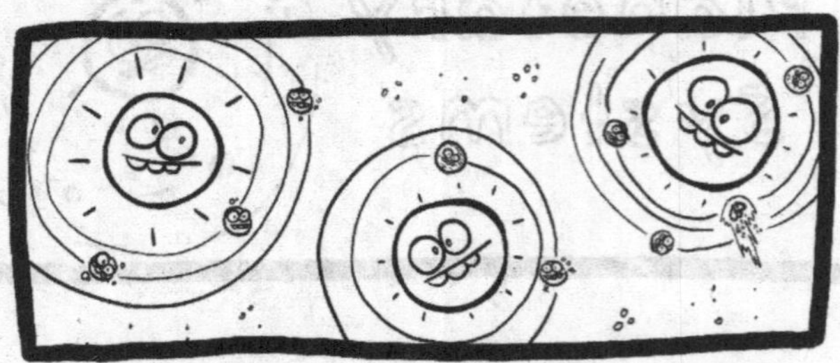

All of that stuff is called a planetary system (the Solar System is a planetary system, too).

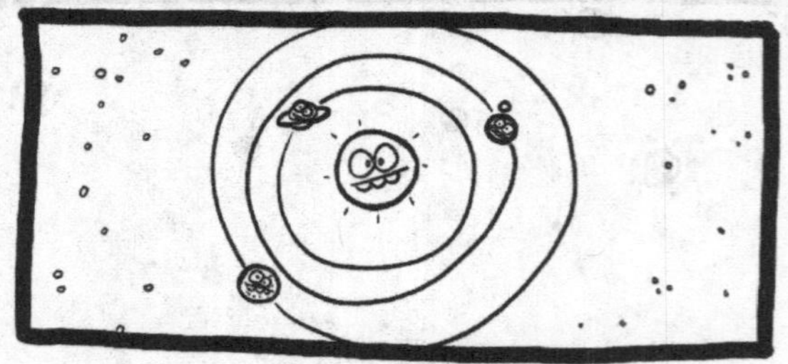

Any planet in one of those other planetary systems is called an exoplanet.

Why Stars Are Hot And Bright

There's a few things you need to know about stars...

There are lots of different stars out there in the universe...

And they are made of tiny things called atoms.

Pretty much everything you see everywhere is made of different types of atoms.

Why Stars Are Hot And Bright

Stars are full of this one type of atom at the start of their lives...

It's called hydrogen.

All stars have a middle bit called a core...

And in that core there are lots of hydrogen atoms getting really squished together.

Hydrogen atoms do something special when they get really squished together...

They fuse (that's when they join up) and become a new atom called helium.

Helium shoots out an energy wave when it first gets made...

This means there is lots of energy shooting out from the helium atoms in a star's core.

That energy moves out of the core and pumps into the rest of the star...

Which makes it really hot and really bright.

The universe has lots of really big gas clouds in it. They're called nebulae (when there's one it's called a nebula)

How Stars Start Out

Inside the nebulae are smaller clouds of gas...

Even though they have hardly any mass, those smaller clouds of gas make gravity...

That gravity makes those small gas clouds pull together...

And make bigger clouds of gas...

How Stars Start Out

Those bigger clouds of gas get pulled together...

And make really big clouds of gas.

How Stars Start Out

Those really big clouds of gas have the mass to make really strong gravity...

And that gravity makes the hydrogen atoms in those really big clouds squish together...

And they fuse to become helium atoms and send out waves of energy...

The energy shoots out of the core of the really, really big gas clouds...

And they become this hot, bright, squiggly, swirly type of star called a protostar...

How Stars Start Out

After a while protostars pull together from their really strong gravity...

And become a round type of star called a main sequence star (like the Sun).

Lots of main sequence stars are about the same size as the Sun...

Lots of main sequence stars are much smaller than the Sun...

And lots of main sequence stars are much bigger than the Sun.

Brown Dwarfs

Brown dwarfs are planets which are really big and really hot...

They have really strong gravity...

But it's not strong enough to fuse their hydrogen atoms...

So instead of becoming a star, brown dwarfs stay stuck as really hot planets.

And they go around a star instead of becoming a star themselves.

All stars have strong gravity pulling in and strong energy pushing out.

The energy pushing out comes from hydrogen fusing in the core (you knew that already).

But eventually a core will run out of hydrogen and it's only got helium left in it...

Once this has happened the star is going to die.

When Stars Die

How a star dies depends on if it is smaller than the Sun, about the same size or way bigger.

When a Sun-sized star runs out of hydrogen, fusing stops and the core shrinks...

That shrinking causes the plasma around the core to grow bigger and bigger...

The star is a lot bigger than it used to be and it is now a red giant.

When Sun Sized Stars Die

After a while the red giant cools down and all that plasma around the core loses its shape...

Breaks away...

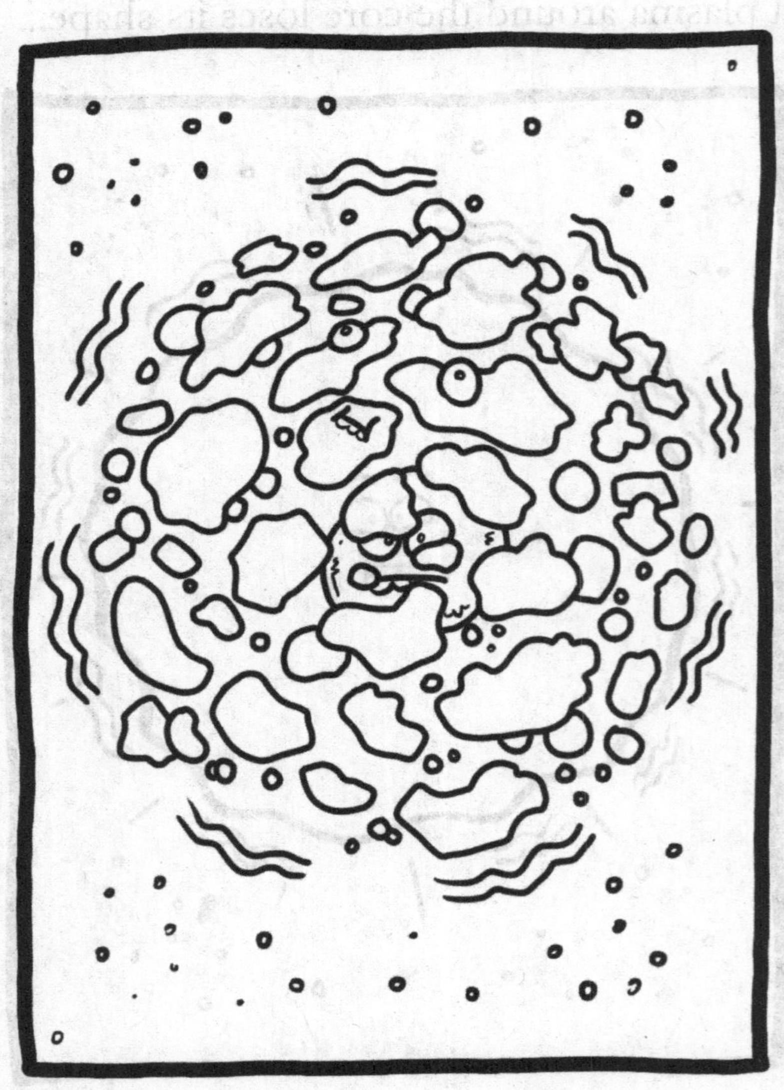

When Sun Sized Stars Die

Becomes a nebula...

And then disappears, leaving just the core.

That core is called a white dwarf.

We're pretty sure that after a long time the white dwarf will stop making light...

And it will stop making heat...

And become this thing called a black dwarf.

When Small Stars Die

Stars that are smaller than the Sun are red dwarfs. They are pretty hot and pretty bright.

But because they're not that hot or bright, red dwarfs don't use up much of their hydrogen.

So they live a really long time, way longer than lots of other things in the universe...

In fact, nobody has seen a red dwarf run out of hydrogen and probably nobody ever will.

When the hydrogen in the core of a big star runs out, the big star just fuses helium...

That's because big stars make gravity that is strong enough to squish helium atoms...

And it heats up a layer of plasma around the core.

The heat around the core is so strong that it makes hydrogen fuse.

All of the hydrogen and helium fusing makes so much energy...

That it causes the plasma to grow and that makes the whole star get bigger...

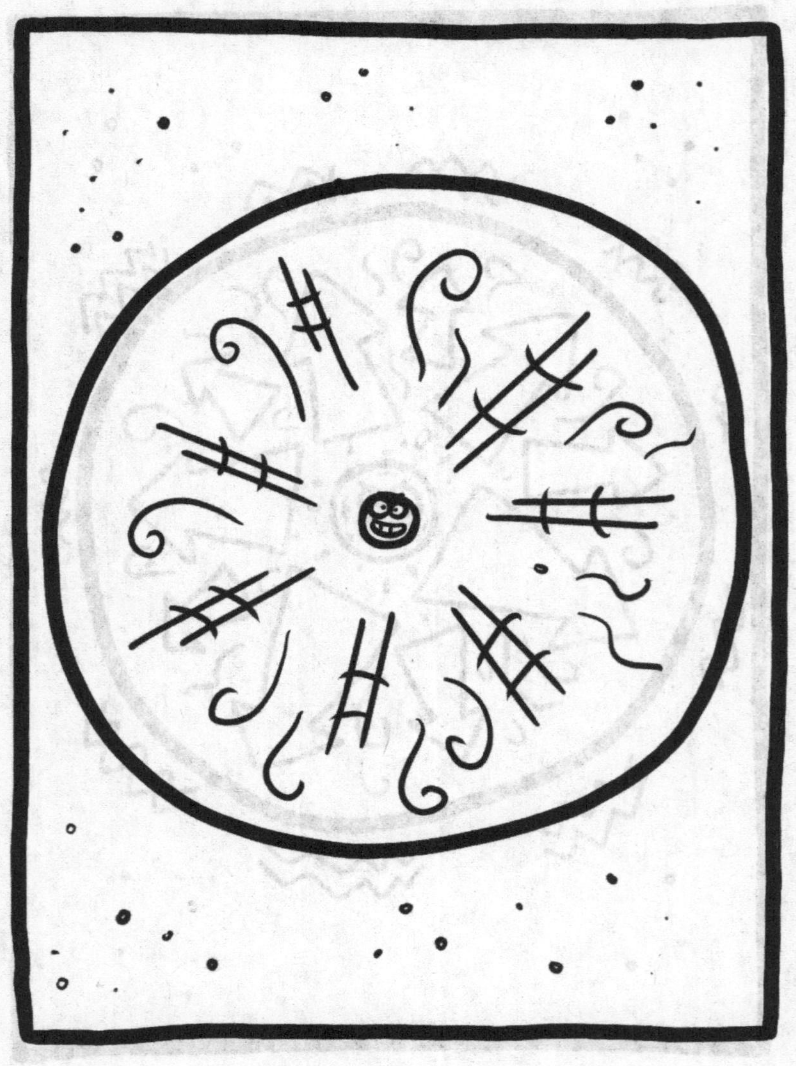

And it becomes a red SUPERgiant.

You might be thinking this all sounds great for the red supergiant so far...

But there's a problem.

And it's happening in the core.

The fusion in the core is making lots of different, brand new atoms and nearly all of them are pumping out energy.

Except this atom. This atom is called iron.

Iron doesn't let out energy when it fuses.

So every time the core makes iron, it pushes out less energy.

Eventually the core builds up lots of iron...

And it doesn't push out enough energy...

To push back against the Sun's gravity.

And at that point the Sun's gravity 'wins'.

It crunches the red supergiant's plasma in on its core.

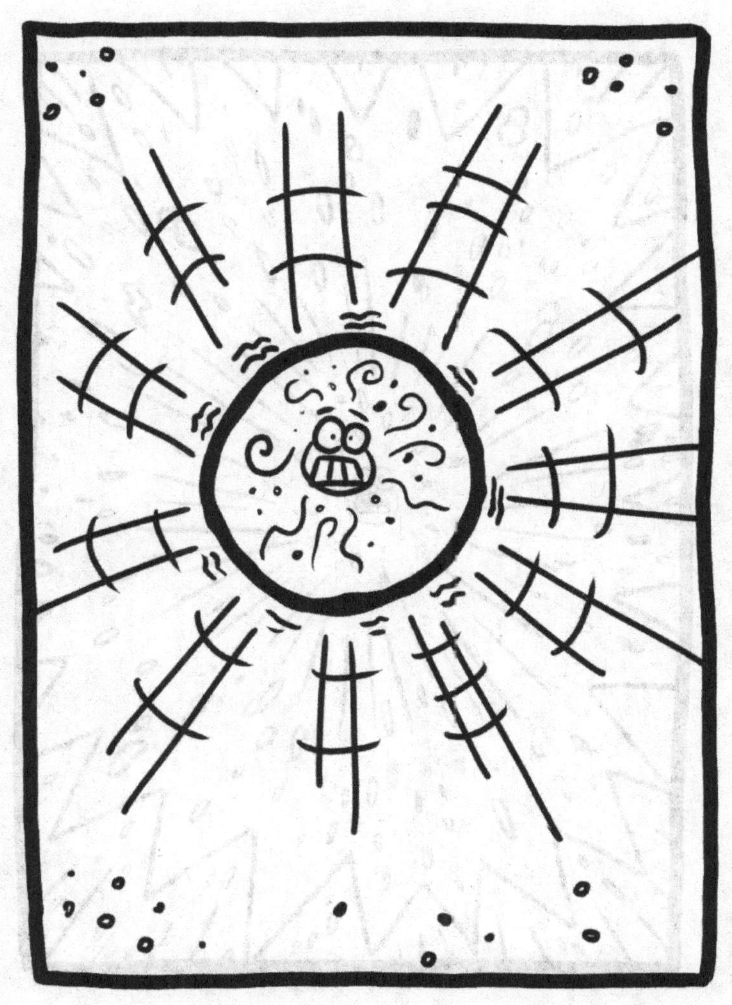

And once the outer plasma is squished into the core it sets off a huge explosion called a supernova.

After the supernova there's a nebula left behind, made of all the bits that blew up.

And then after the nebula clears away there could be nothing left behind...

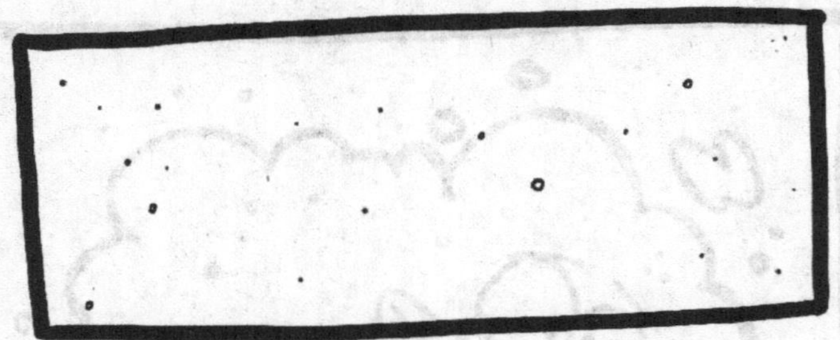

Or this thing called a neutron star...

Or this other thing called a black hole.

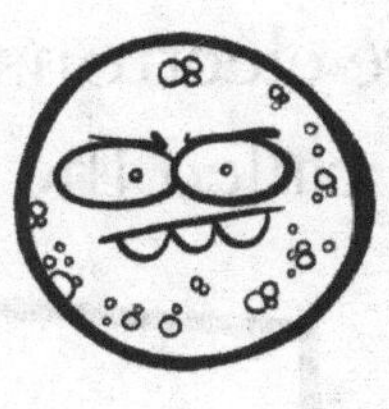

All atoms have three main bits in them...

There are electrons going around the middle bit which is called the nucleus.

And if you look inside the nucleus you find these two things called neutrons and protons.

When a red supergiant collapses, the gravity is so strong it even crushes the atoms in the core.

The crush is so strong it pushes the electrons that were originally outside the nucleus into it.

Then it squishes the electrons up against the protons in the nucleus...

And they all squish together and become...

Neutrons.

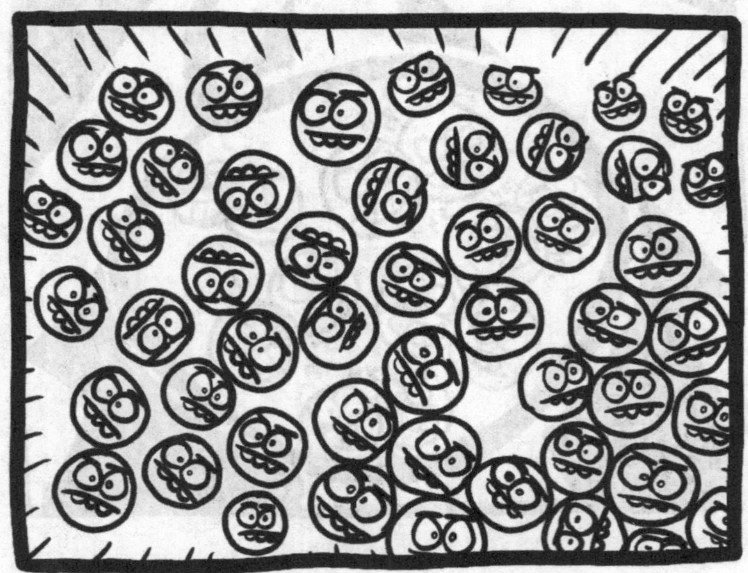

Neutron Stars

When the old core only has neutrons it has become a neutron star.

Neutron stars are much smaller than stars...

But they can have much more mass.

Which means they have really strong gravity.

Neutron stars also spin really, really fast.

And they're really, really hot.

Black holes start with a core being crushed by really strong gravity...

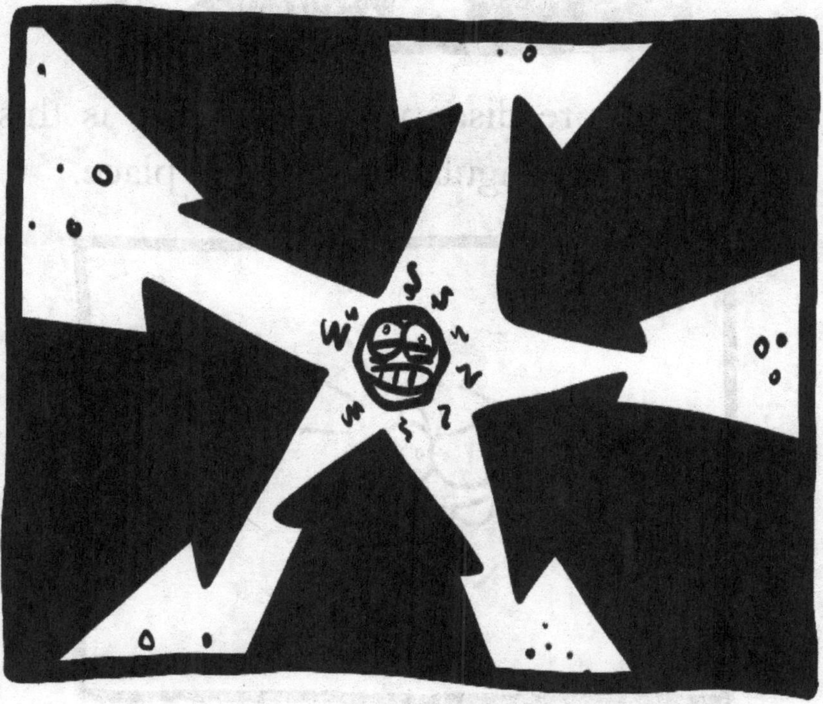

And this time the gravity keeps crushing the core down and down and down...

Until the core disappears and there is this thing called a singularity left in its place.

A singularity is a spot in the universe that makes really, really, really strong gravity.

Its gravity is so strong it pulls in everything near it. It even pulls in light.

Soon a big ball of darkness will grow around the singularity. It's called a black hole.

Black holes have the strongest gravity in the universe. They pull in everything...

And the more black holes pull in the bigger they get.

Black holes can even pull in other black holes...

A black hole can start when a neutron star pulls in lots of stuff with its gravity...

And builds up so much mass that it gets really, really, really strong gravity...

Collapses and turns into a singularity...

And then a black hole.

Black holes can also start when two neutron stars run into each other...

And make a huge explosion called a kilonova.

The gravity from each neutron star can join up to crush them in on themselves...

And cause a singularity...

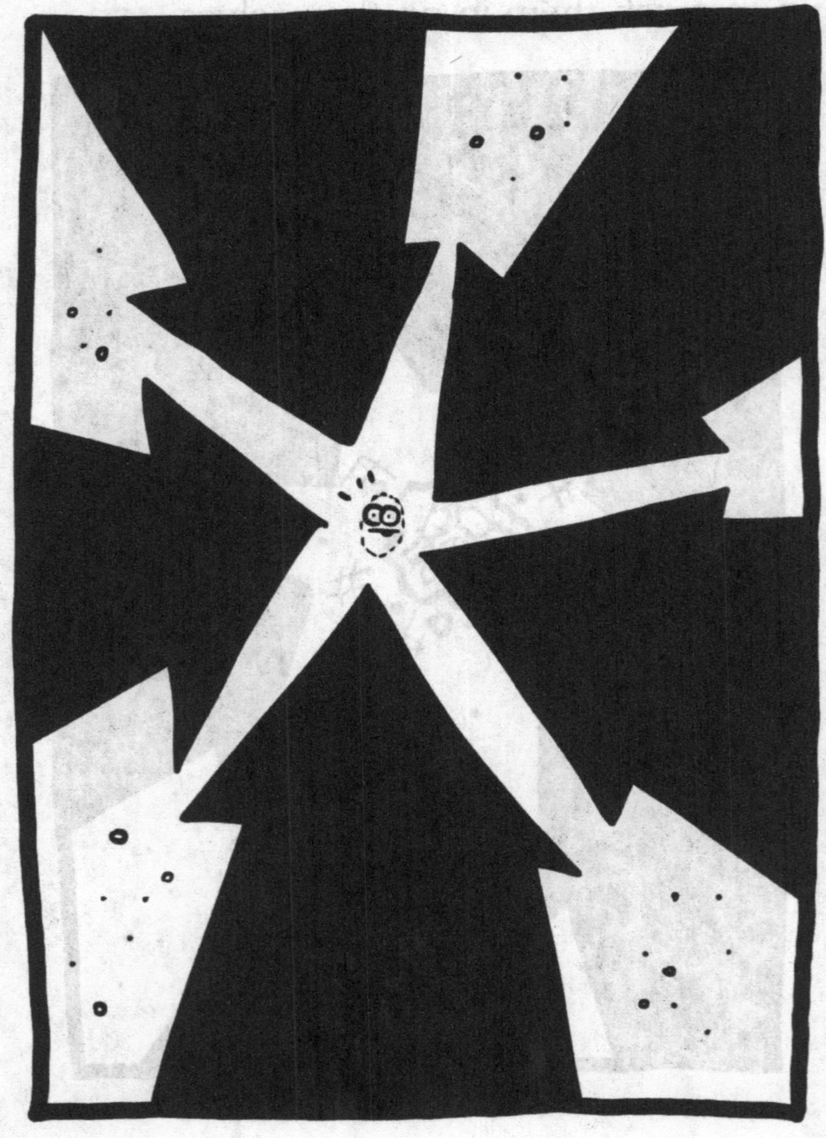

Which causes a black hole.

You're probably worried that black holes will end up pulling everything in...

Until the whole universe is nothing but one big black hole. That's fair enough.

But don't worry. It won't happen.

Firstly, black holes can only pull in stuff that is inside the reach of their gravity.

And secondly, we're pretty sure black holes are dissolving away really slowly...

And eventually they will just disappear.

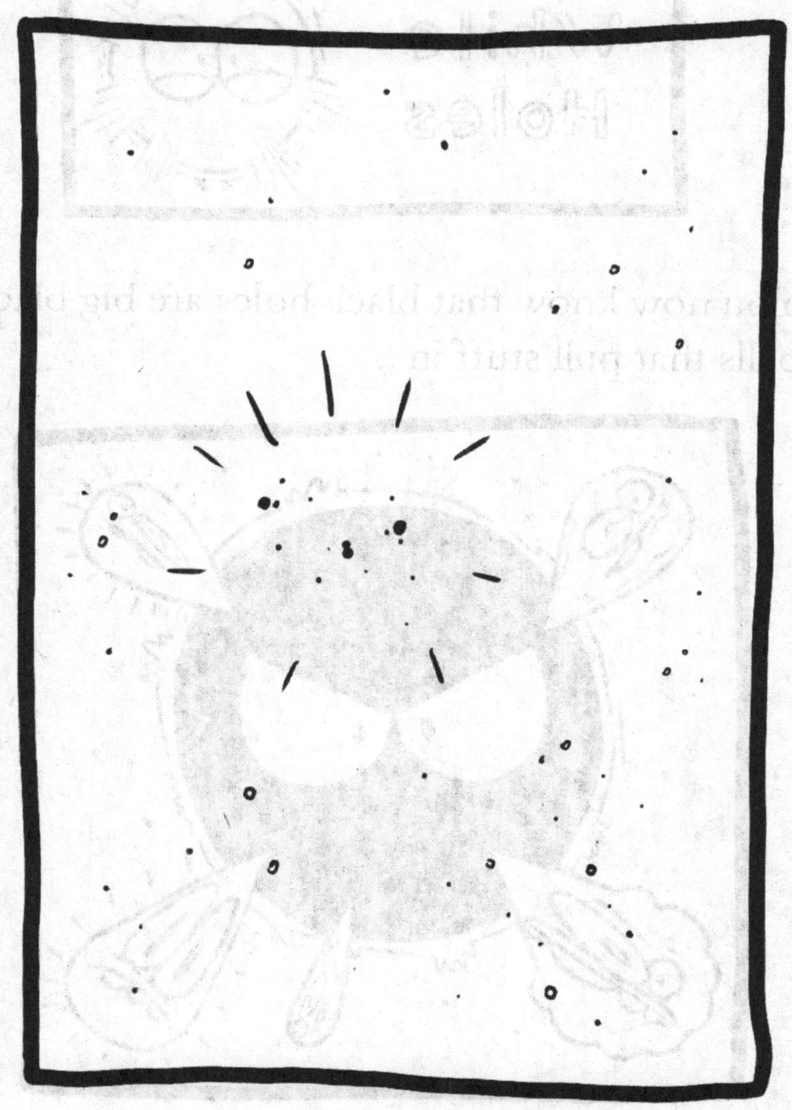

You now know that black holes are big black balls that pull stuff in...

Well, we think that white holes are bright white balls that shoot stuff out.

Black holes last a really, really long time.

But white holes probably just pop up, shoot stuff out.

And disappear very quickly.

Nebulae are huge clouds going around in space made of dust, gas or plasma.

There are different types of nebulae...

This is a planetary nebula. It's called a planetary nebula because it's round like a planet.

They come from the plasma that a red supergiant shoots out as it is dying.

This is an emission nebula. It is the type of nebula that protostars come from.

When a supernova happens it shoots out lots of stuff into space.

After the supernova is over it leaves behind clouds of gas.

Those clouds of gas are a type of nebula called supernova remnants.

Reflection nebula look really bright and shiny.

That's because they are full of dust that is reflecting the light from bright things nearby.

Dark nebula look really dark.

And that's because they are blocking the light from something bright nearby.

The Solar System is full of stuff orbiting the Sun.

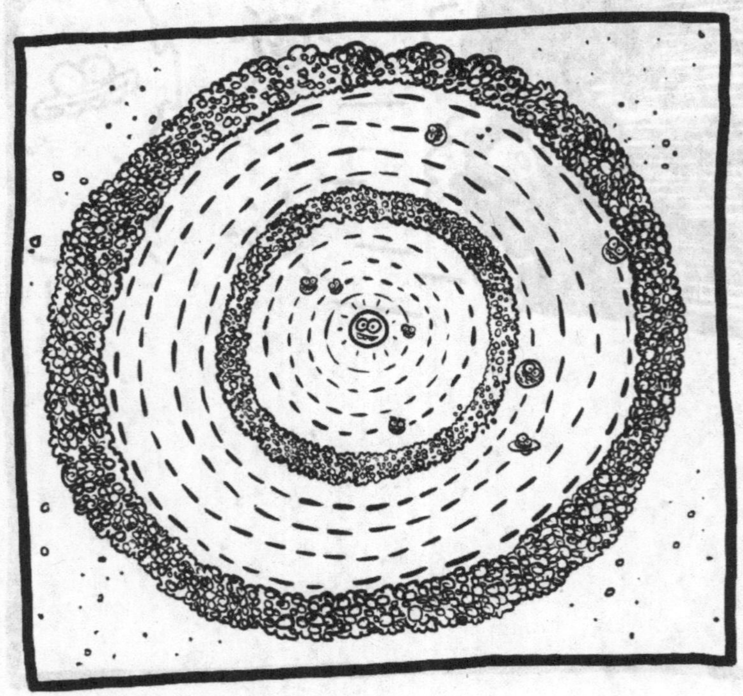

But the Solar System itself is moving too.

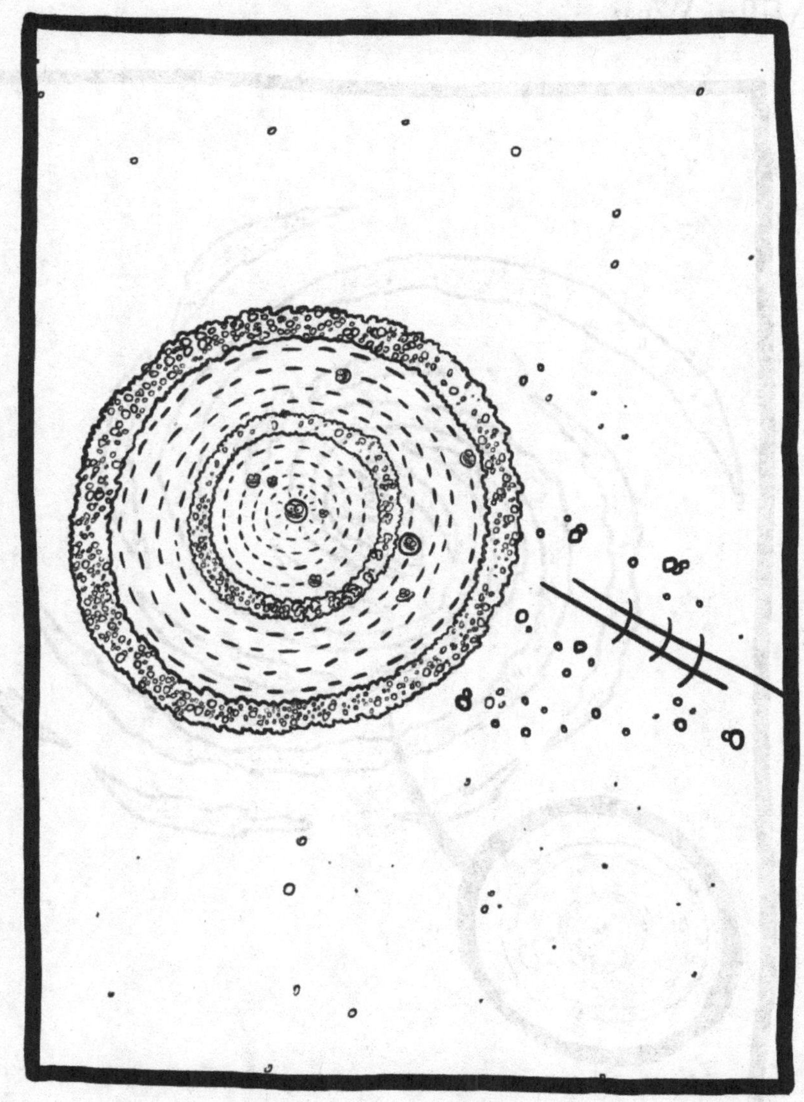

It's moving around in a galaxy called the Milky Way.

Galaxies

Galaxies are made up of dust, gas and planetary systems spinning around together.

And in the middle of every galaxy is a really, really, really big black hole.

There are three main types of galaxies...

Spiral galaxies have a bunch of arm shapes spinning around their middle bit.

Elliptical galaxies are shaped like a glowing fried egg on a pan.

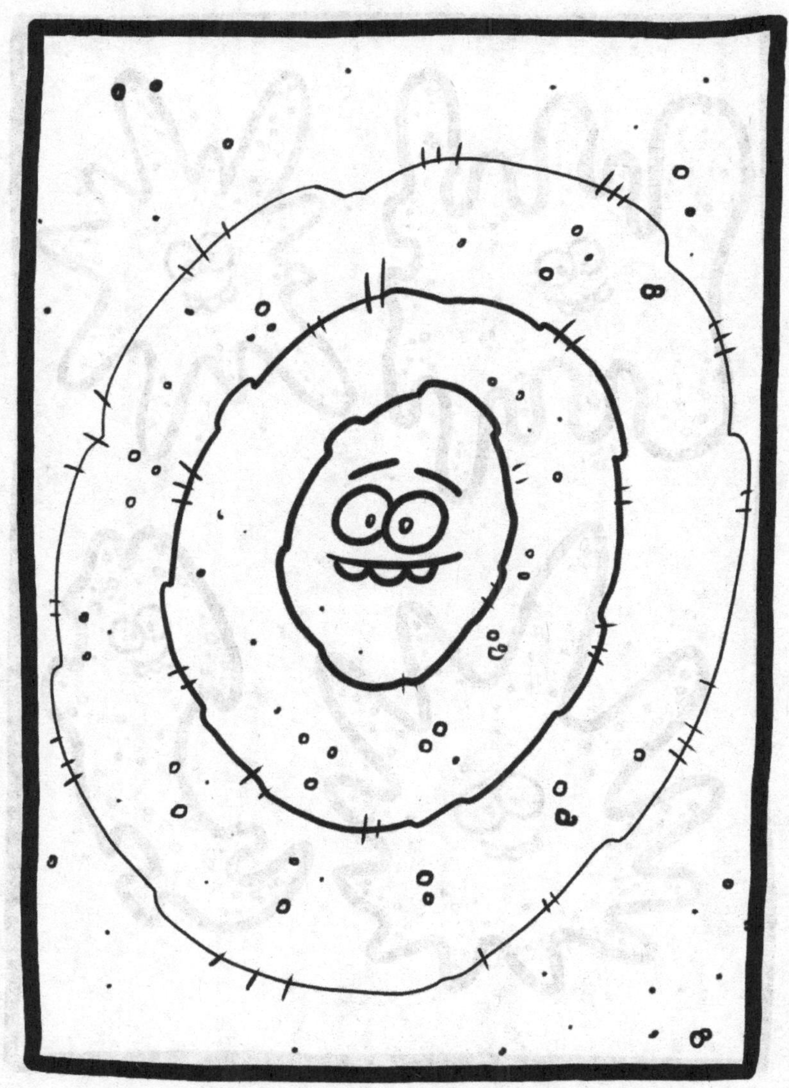

Irregular galaxies are the ones that aren't spiral or elliptical. They can be lots of shapes.

Galaxies

Galaxies are moving around space, too.

Sometimes they even run into each other.

When this happens they can join up to make one big new galaxy...

Or they can just pass through each other, pop out the other side and keep going...

Galaxies can also spin around together in a galaxy cluster.

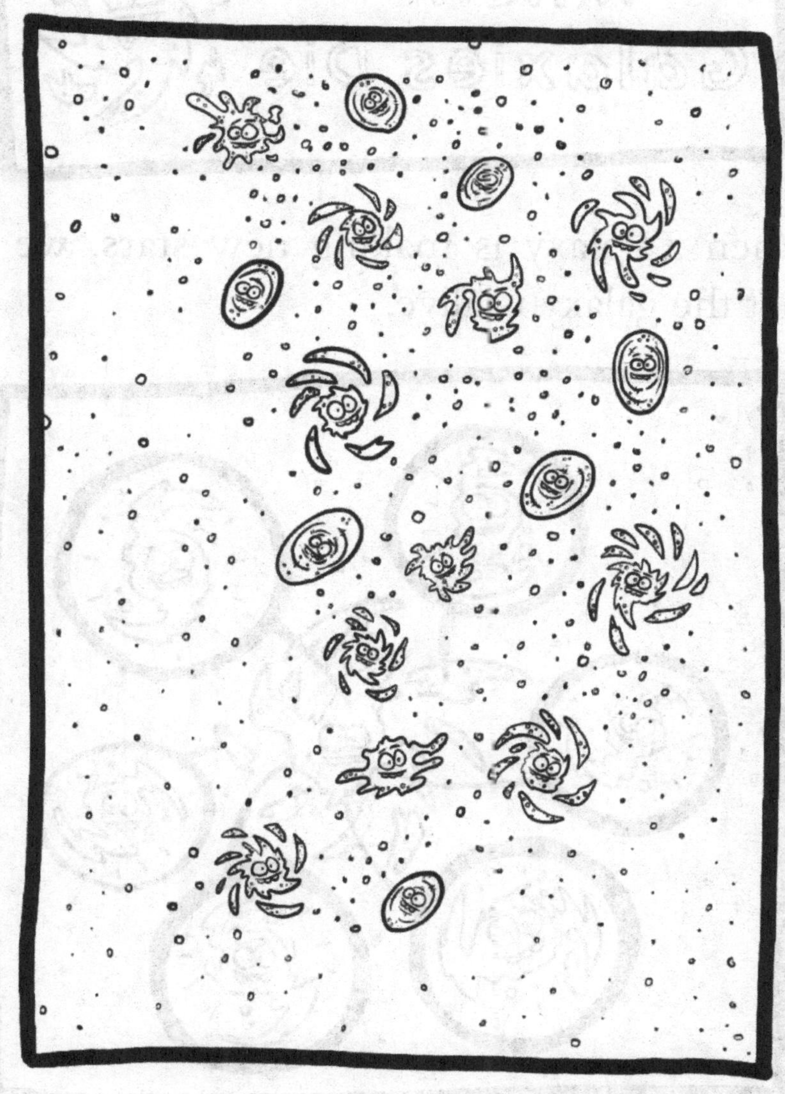

When Galaxies Die

When a galaxy is making new stars, we say that the galaxy is alive.

To make new stars a galaxy needs lots of emission nebulae full of hydrogen atoms.

But after a very long time a galaxy will run out of the hydrogen it uses to make new stars.

When Galaxies Die

When galaxies stop making new stars they are called dead galaxies.

Dead galaxies slowly shrink down...

And end up looking like squished red blobs...

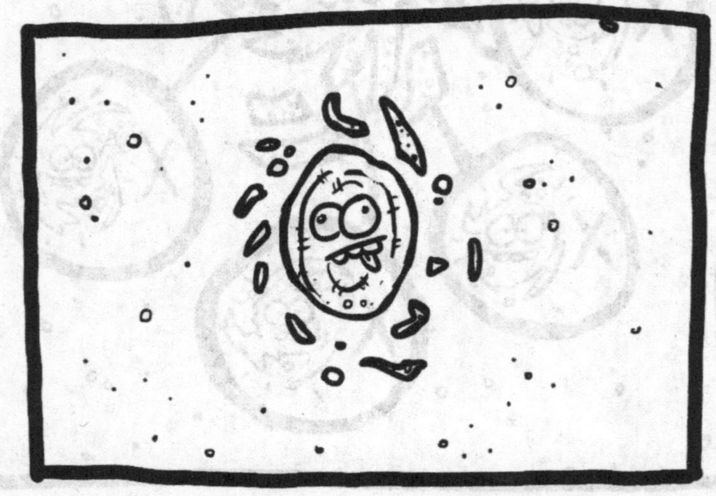

But if a dead galaxy goes around...

And comes across a smaller living galaxy...

The stronger gravity of the dead galaxy pulls away hydrogen gas from the smaller galaxy...

And the dead galaxy uses that gas to make stars. This brings the dead galaxy back to life.

Galaxies that used to be dead which come back to life are called zombie galaxies.

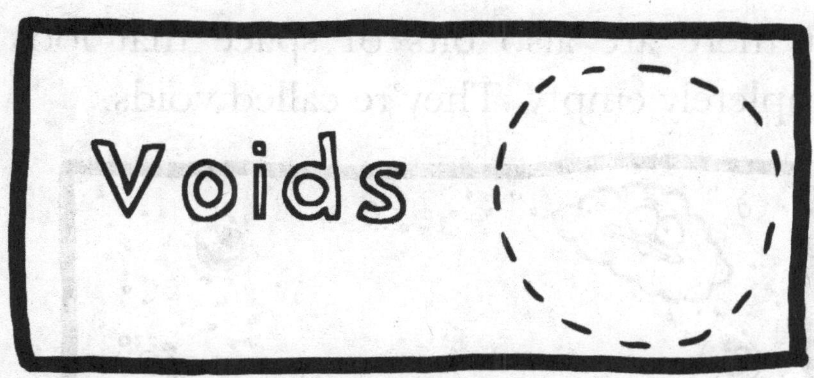

You can see plenty of stuff in space...

But there are also bits of space that look completely empty. They're called voids.

Voids

Voids look empty, but they're probably full of stuff called dark matter and dark energy.

Matter is the name for the stuff we can see or touch (called solids, liquids, gas and plasma)

Solids keep their shape like rocks or tables. Liquids are drippy and splashy.

Gases are whooshy and puffy. Plasma are like really hot gases and found in stars and nebula.

Dark matter is probably nothing like matter. We don't even know what it's made of.

Or what it looks like (it's invisible)...

And when dark matter comes across something made of matter in the universe...

They probably pass straight through each other.

We're pretty sure that dark matter has gravity.

And there's probably a lot of it in the universe...

In fact, galaxies are probably held together by all the gravity that dark matter makes.

This is energy. Energy does lots of things.

If something has energy...

It can get hotter...

Or it can move.

This is dark energy. It makes a type of gravity that pushes stuff away instead of pulls it in.

And it pops up from empty space.

Dark energy has filled up all the bits of the universe that look empty...

And all that dark energy is making lots and lots of gravity that pushes out...

Which means everything in the universe is being pushed away from everything else.

Dark Energy

Dark energy is pushing out all over the universe...

It's even making the universe itself grow...

But it's not just making the universe grow. It's making it grow faster and faster...

That's because every time the universe grows, it makes more empty space...

More empty space means more dark energy popping up and pushing out...

Which just makes the universe grow even faster and faster...

Spacetime

Spacetime is like a map of the universe. It sounds fancy but it's not that fancy.

You already know that everything in the universe is always moving around.

So something could be here now...

But the next time you look it's gone.

So to find stuff in the universe, you don't just think about WHERE it is...

But WHEN it is there, too...

This means time and space are connected.

Space and time together make spacetime: the universe's moving map.

The lines of spacetime show how space and time move.

When something with mass is put in spacetime...

It dents space and changes how it moves.

When other stuff with less mass gets near the dent...

It can fall in...

And get stuck.

Or if it is going really, really fast when it falls into the dent...

It can shoot in...

And shoot out the other side.

We can't really see these lines and dents when we look at the universe...

We see them as gravity and orbits and the movement of stuff.

Remember that time flows along the lines of spacetime, too? Of course you do.

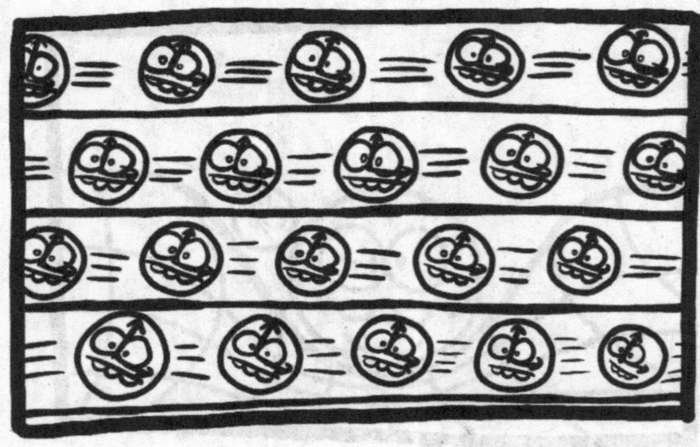

When something puts dents in spacetime it also slows the flow of time...

This means that when you get close to things that have strong gravity in the universe...

Time moves slower for you.

And when you get further away from things that have strong gravity...

Time moves faster for you.

Of course, you wouldn't notice time going faster or slower...

It would feel like time was moving at the same speed...

No matter where you were.

This all means that if we got you together with someone who was the same age as you...

Then shot you off into space...

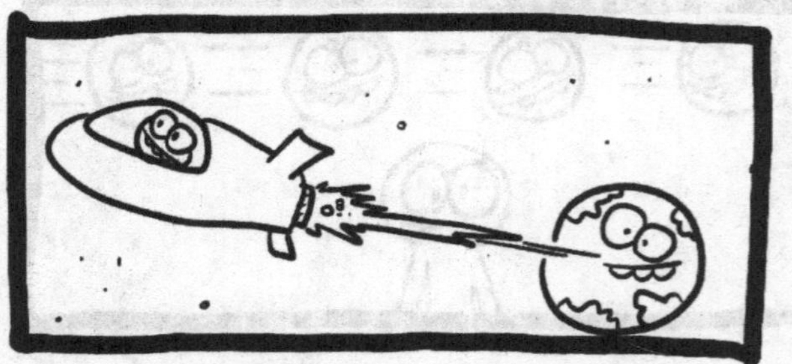

And left the other person on Earth...

Then sent you back to Earth after a while...

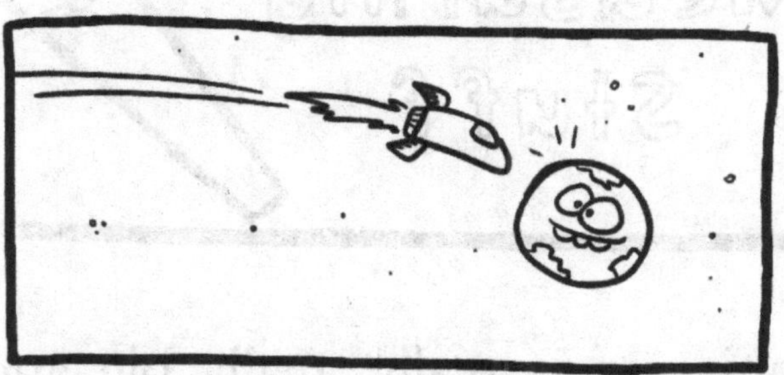

You would be older than the person that used to be the same age as you.

And that is because time would go faster for you in space than the other person on Earth.

The universe is really, really big and the distances in it are really, really big too.

So to measure stuff in the universe we use astronomical units, lightyears and parsecs.

This is an astronomical unit (also called an AU). We use it to measure stuff in space.

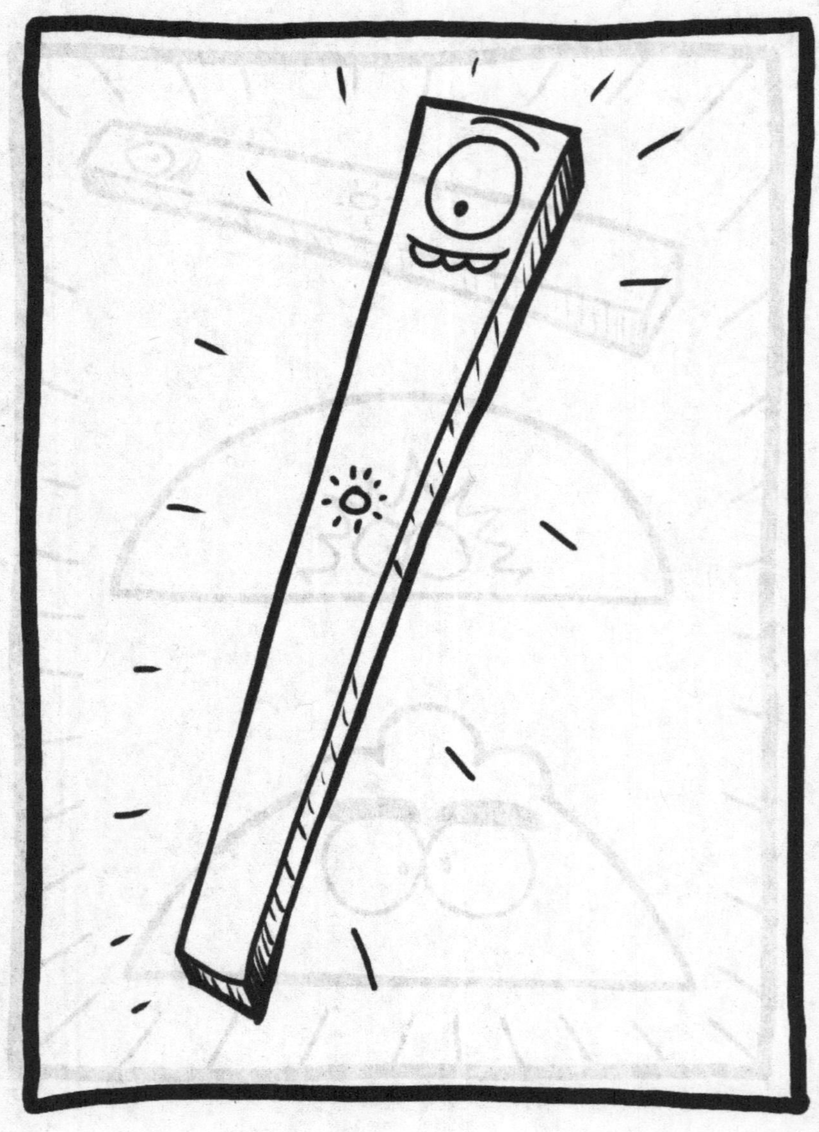

Its length is the distance from Earth to the Sun (about 149 and a half million kilometres).

You can use just part of an astronomical unit to measure pretty small distances...

Or you can use lots of astronomical units to measure big distances between stuff.

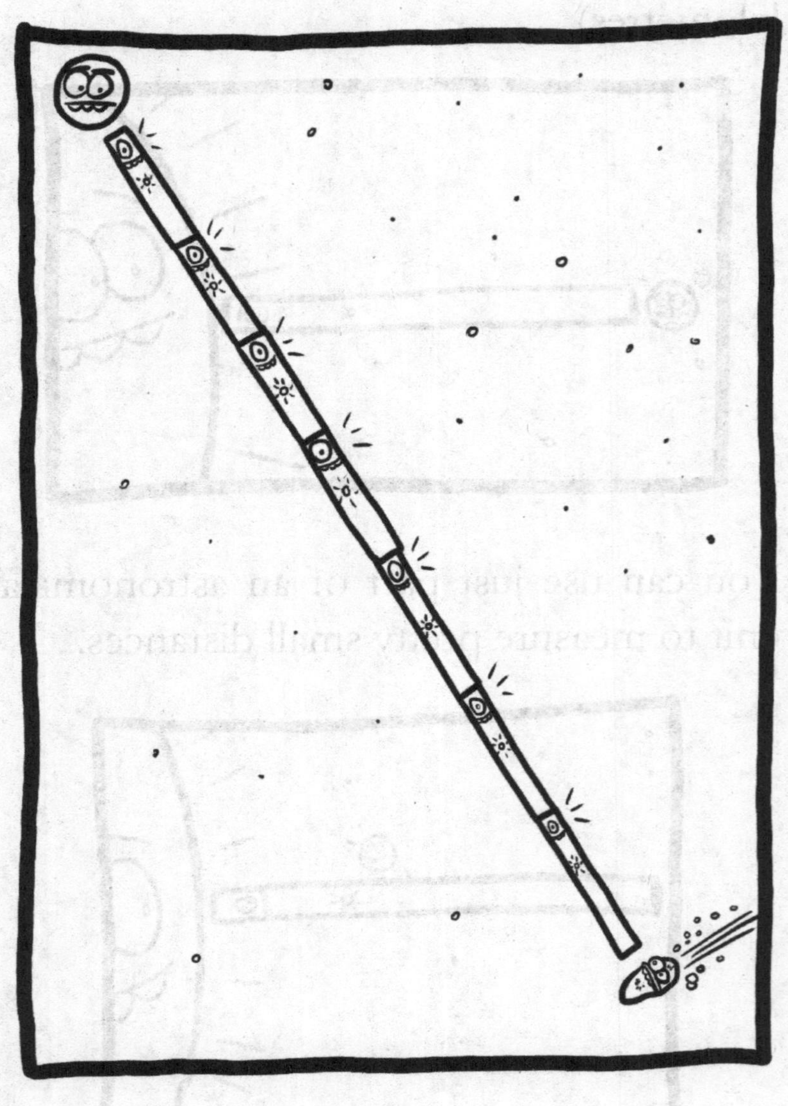

This is light. Light moves really fast in space.

In just a second of moving through space, light has covered about 300 000 kilometres.

In a year, light has moved about 9 and a half trillion kilometres through space.

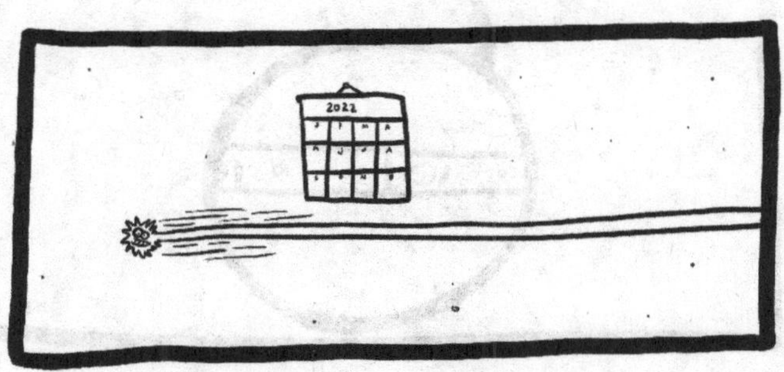

A lightyear is equal to the distance that light moves through space in a year.

It is a bit over sixty three thousand astronomical units long.

Lightyears can be used to measure big distances between planetary systems.

You can even use thousands of them to measure really, really big distances.

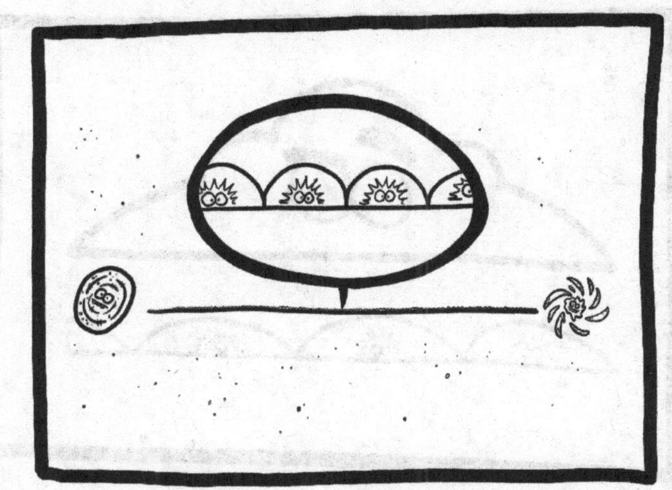

This is a parsec. It can also be used to measure stuff in space.

A parsec is three and a bit lightyears long.

Parsecs are used to measure really big distances between stuff in the universe.

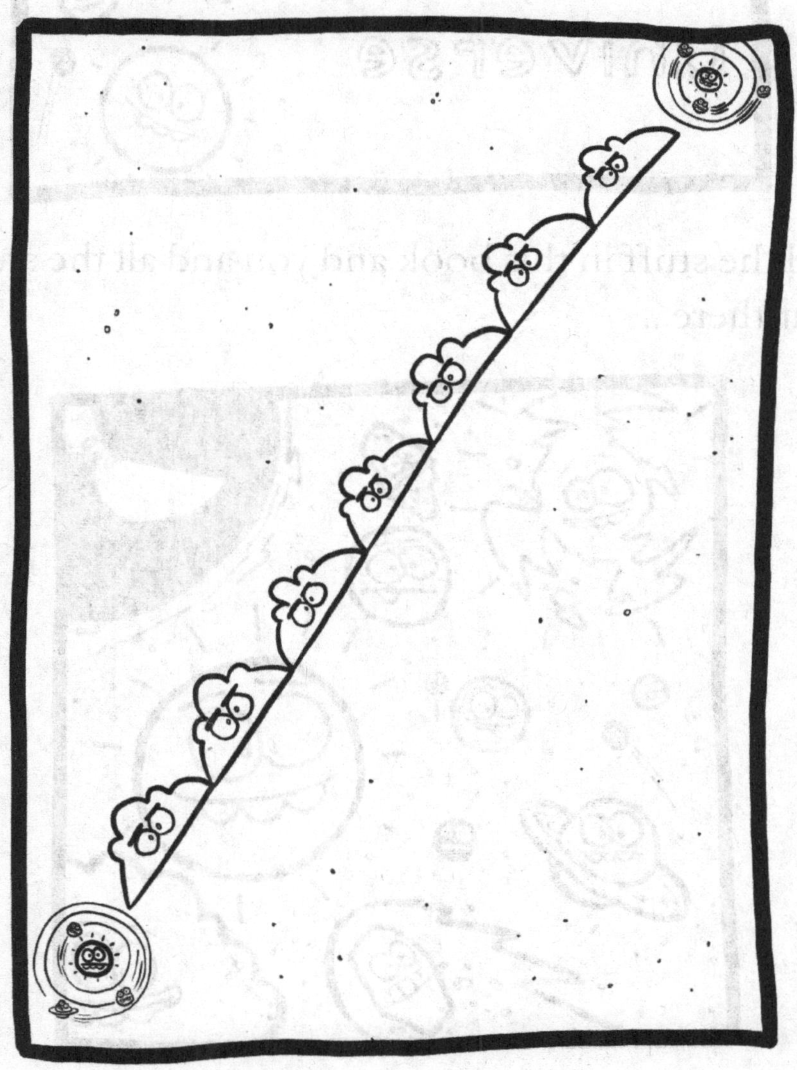

All the stuff in this book and you and all the stuff out there...

The Universe

Is part of the universe.

About 5 percent of the universe is made of matter, 25 percent is dark matter and 70 percent is dark energy.

The Universe

Right now we think that the universe probably goes on and on straight in every direction...

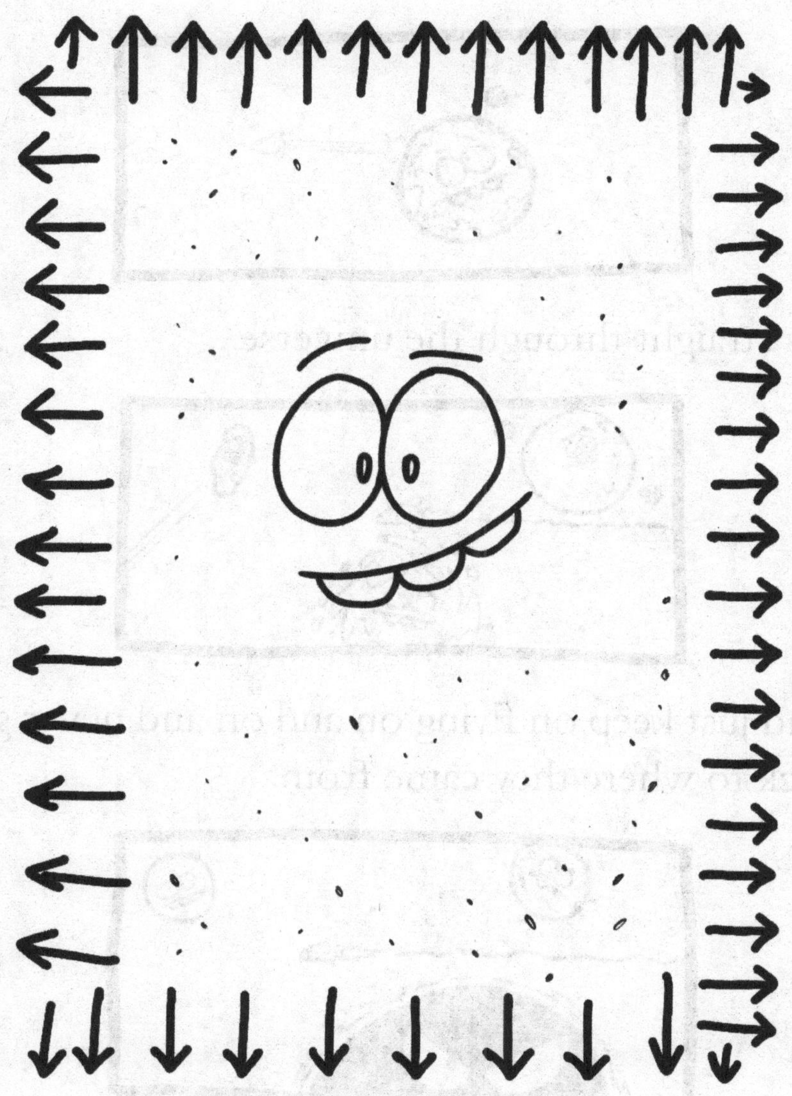

That means we think someone could leave Earth...

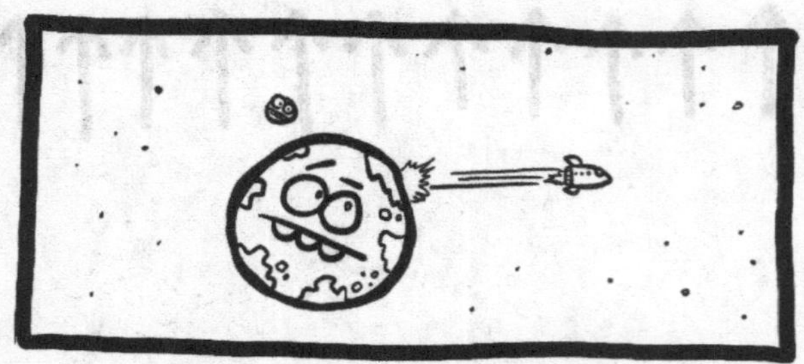

Fly straight through the universe...

And just keep on flying on and on and never get back to where they came from.

But the bit of the universe we can see is probably only a piece of a much, much bigger universe.

And maybe if we saw all of this much, much bigger universe we'd see that it is actually round.

Being in a round universe would mean that if someone left Earth...

They could fly straight through the universe...

And after a really long time get back to Earth.

If that is weird to picture, try this:

Picture yourself flying out into the universe on a line that you think is perfectly straight...

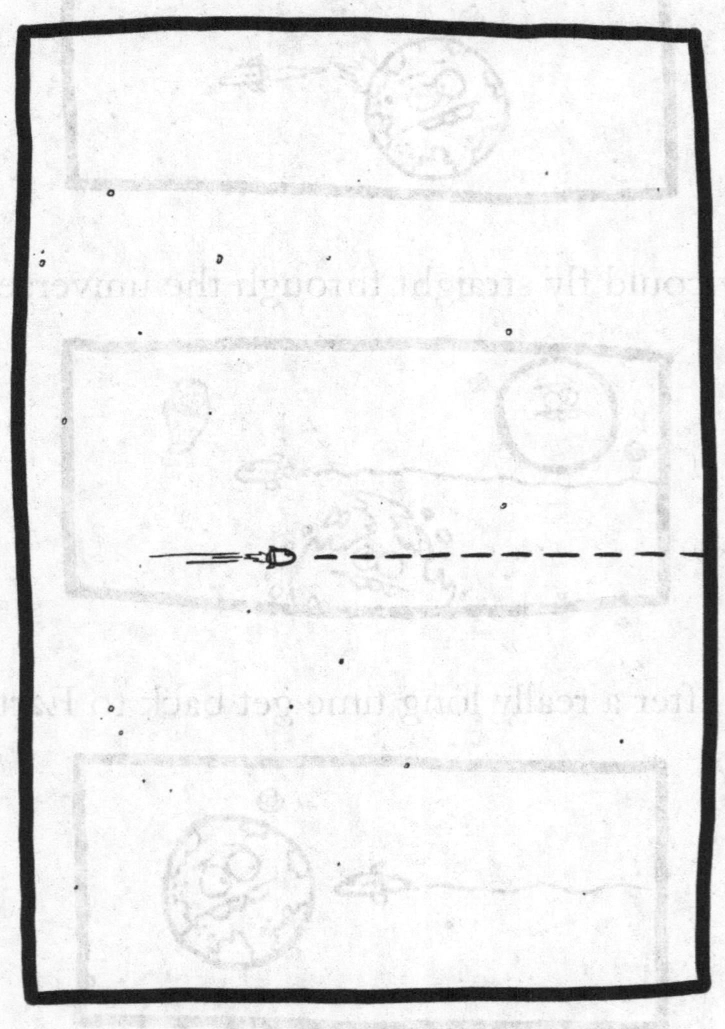

But that 'straight' line you're on is actually curving with the curve of the universe...

So, even though you think you're going straight, you're actually going round...

And round...

And round until you end up back where you started.

To see what happened at the start of the universe you need to meet some of the little stuff...

Let's start with an atom. You know what an atom is. You're an atom expert.

You know that atoms have a middle bit called a nucleus...

And electrons going around the nucleus.

Electrons are a type of thing called a lepton.

There are lots of different types of leptons.

Now if you popped open a nucleus you would find protons and neutrons...

Neutrons and protons are both hadrons. There are lots of different hadrons out there.

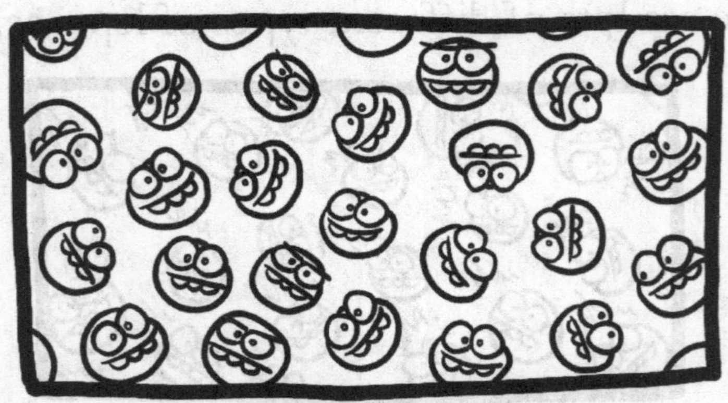

And if you opened up protons and neutrons or any type of hadron...

You'd find quarks.

Quarks are really, really small and they always go around in groups of three.

This is light...

Light is made of these little things all stuck together. They're called photons.

Photons are the last thing you need to know about to see the start of the universe.

All the stuff in the universe and all the space in the universe and even the universe itself...

Started out squashed into a space smaller than the dot in the middle of this page...

This squashed up universe was really hot and packed really, really tight with stuff.

Then about 13 and a bit billion years ago, the universe started expanding. It was the Big Bang.

The Start Of The Universe

The universe was expanding really, really fast...

Inside the expanding universe were just photons, leptons and quarks shooting around...

Then the universe got even bigger and cooler...

The quarks joined up...

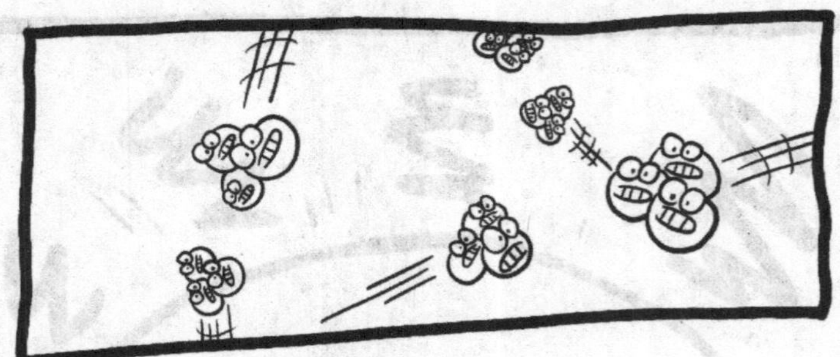

And made hadrons like neutrons...

And protons.

Then the universe got even bigger and cooler...

The neutrons and protons joined up...

And each made nuclei (when you have more than one nucleus, you call them nuclei)...

The Start Of The Universe

Then the universe got bigger and cooler...

The nuclei pulled in the electrons...

And joined them to make atoms...

Most of those atoms were hydrogen atoms...

Lots of hydrogen atoms joined up...

To make nebulae...

There were lots of nebulae in the universe...

They even got together to make the first galaxies.

In the nebulae, protostars formed...

And became main sequence stars...

And planets formed around the main sequence stars from leftover bits of nebulae...

So did asteroids, comets, dwarf planets and lots of the other stuff found in planetary systems...

After some more time and stuff happening the universe turned out the way we see today.

Something else was happening during the Big Bang...

There was lots of matter going around...

But there was also stuff called antimatter going around during the Big Bang, too.

Antimatter

Antimatter acts a lot like matter. But when matter and antimatter touch...

They blow each other up.

And leave nothing behind.

This means if there were equal teams of matter and antimatter at the start of the universe...

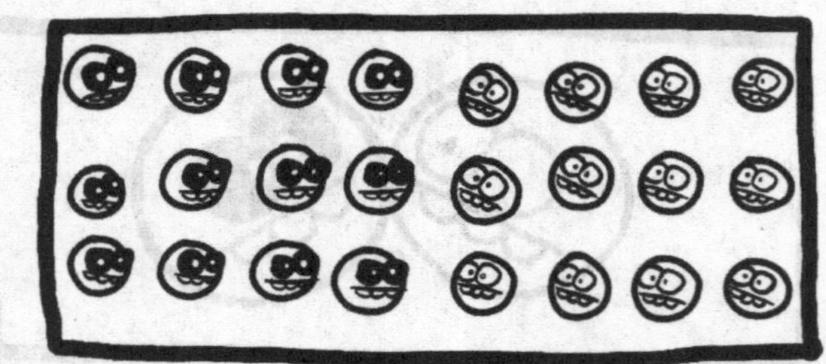

They would have all blown up.

And there'd be nothing left after the Big Bang. No you, no me, no planets, no stars. Nothing.

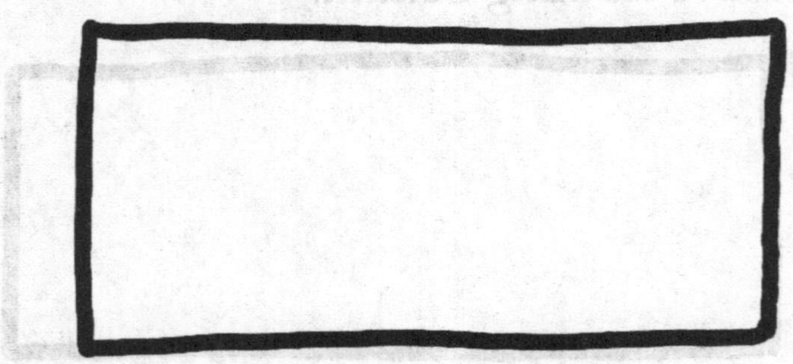

But you may have noticed there IS actually lots of stuff in the universe.

Lots of it is made of matter...

And none of it is made of antimatter.

This could mean there was a little more matter than antimatter during the Big Bang...

So once all the explosions of matter and antimatter were done...

There was that bit of extra matter left over to make all the stuff in the universe.

Or maybe there were equal amounts at the start but some matter and antimatter never blew each other up.

And after all those explosions the left-over matter went on to make stuff in the universe...

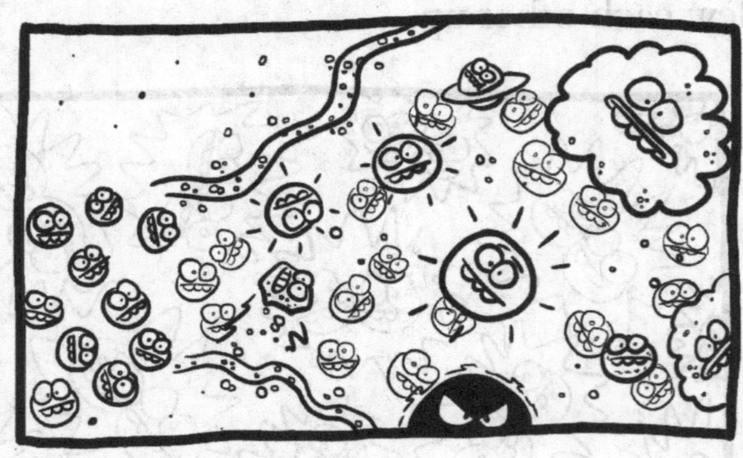

And the left-over antimatter was hidden somewhere in space that we haven't found yet.

The End Of The Universe

In billions of years, the universe will probably end. We don't know how but we can guess...

Maybe everything runs out of heat and it all goes cold...

And dark.

That ending is called Heat Death.

Or maybe the universe ends up with so much dark energy that it pushes everything apart...

The dark energy even pushes apart atoms.

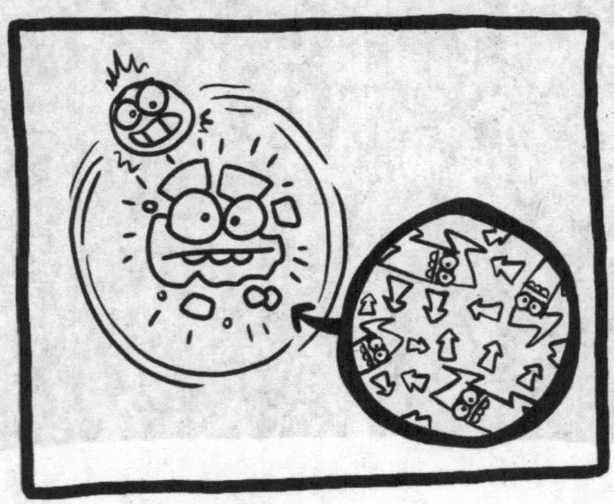

And the universe ends up being full of nothing but ripped-up bits of atoms spread out all over the place. That ending is called the Big Rip.

Or maybe the universe gets really, really, really big...

Then it just squashes back in on itself (this is called the Big Crunch).

And it just keeps getting smaller and smaller...

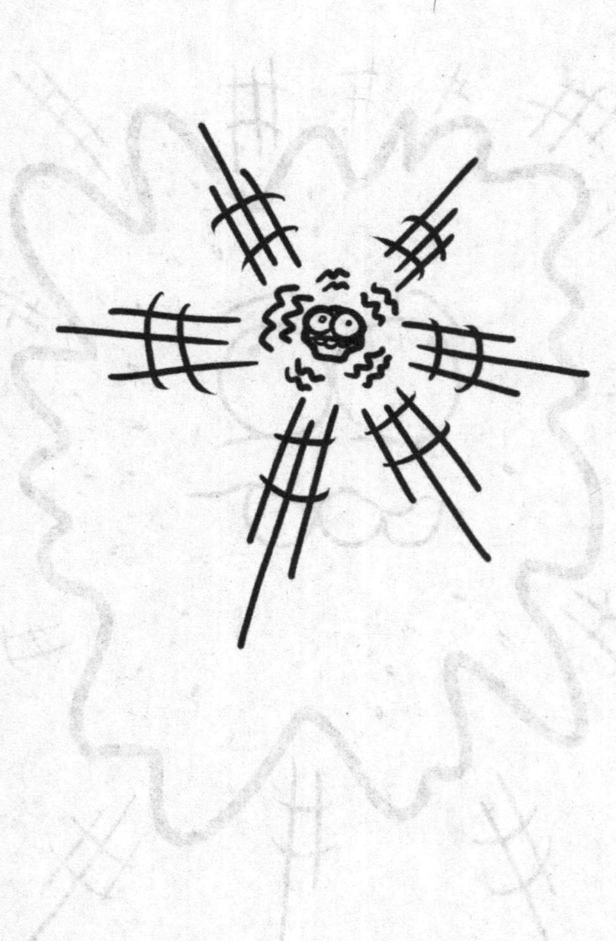

And smaller still.

Until it suddenly bounces back and grows again (This part is called the Big Bounce)

And turns into a new universe.

The Big Bang was when the universe had started to grow from a squashed up dot...

But what first put that squashed up dot there?

What Caused The Big Bang?

Maybe there was nothing before the Big Bang and only the dot was there at the start.

That would mean the Big Bang was the first thing that happened in all of time...

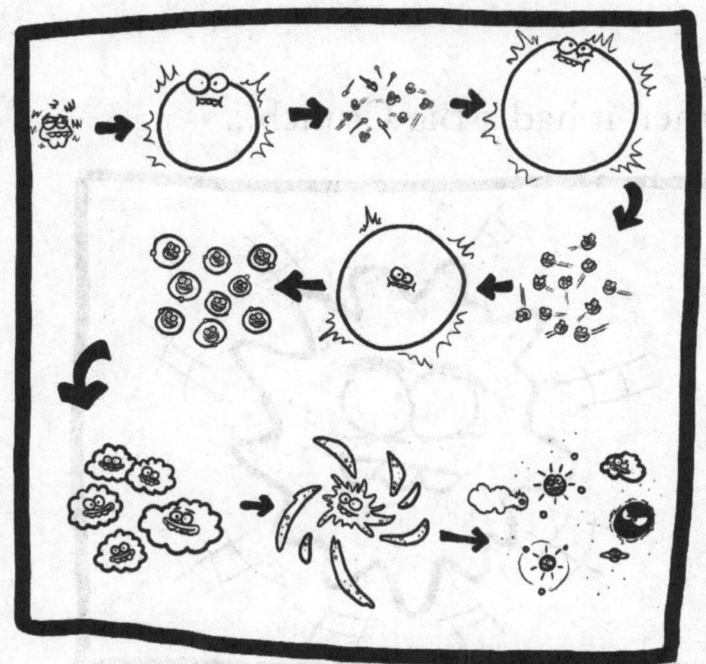

Or maybe there was another universe before the universe we are in...

And then it had a Big Crunch...

What Caused The Big Bang?

Got squished all the way down...

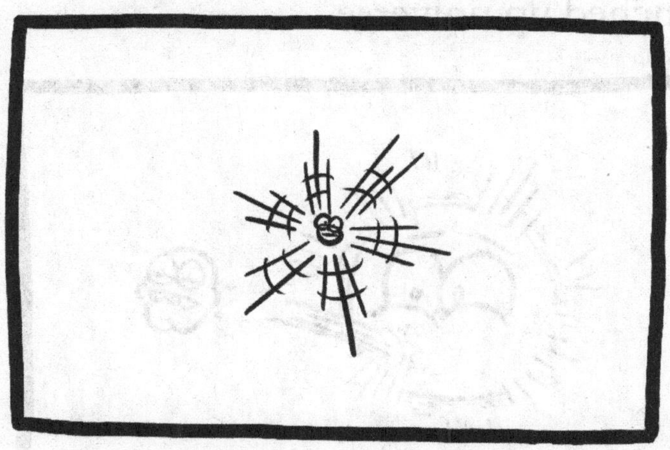

And the Big Bang for our universe got started from that old scrunched up universe.

Maybe a white hole popped up and spat out a scrunched up universe.

And the Big Bang got started from that.

Or maybe there was nothing but energy zapping around...

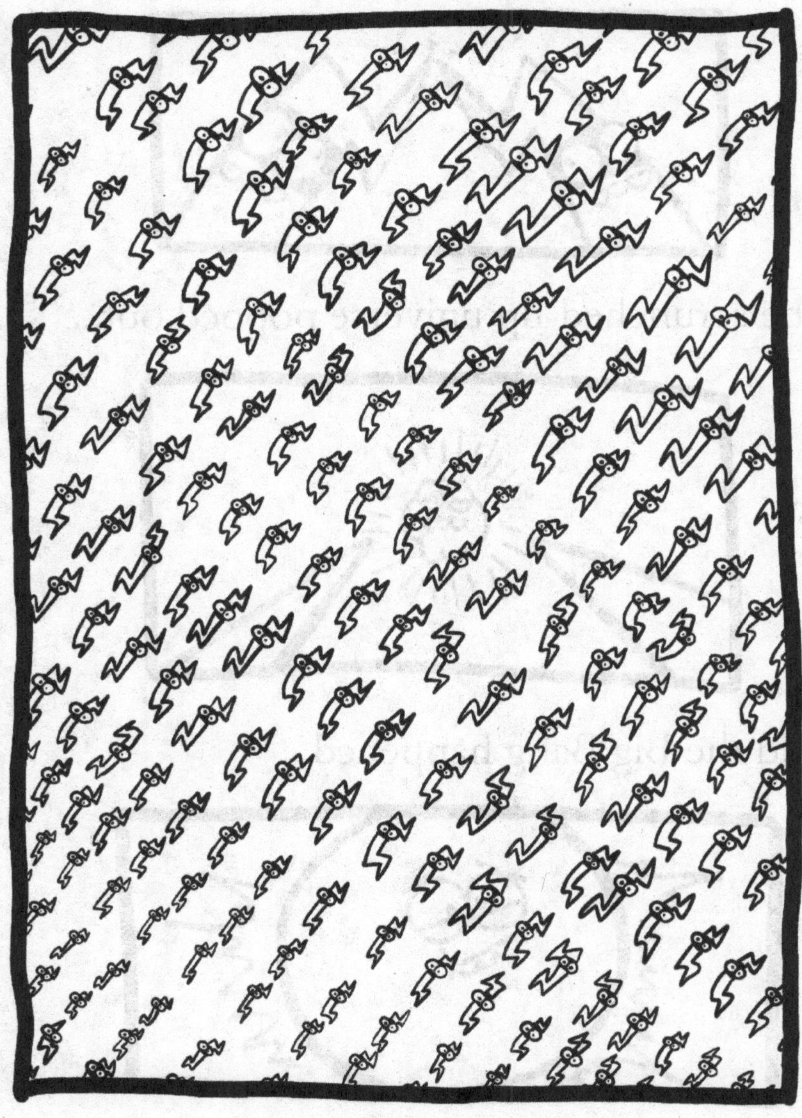

Then some energy zapped off some other energy...

The scrunched-up universe popped out...

And the Big Bang happened...

The universe is the biggest thing we know about...

But we might be in one universe among many universes and all of our universes are in this bigger thing called the multiverse.

Each different universe in the multiverse could be pretty much the same as ours.

Or they could be a little different...

Or really, really different.

Each universe in the multiverse could be the same shape...

Or they could look very different.

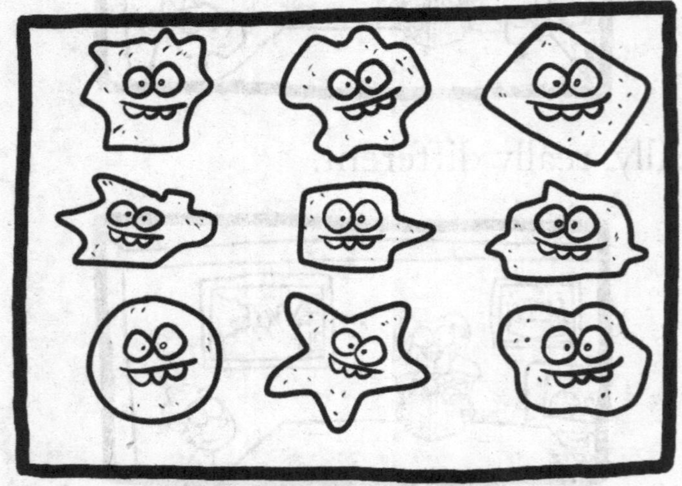

There might be just a few other universes out there in the multiverse...

Or there could be lots and lots and lots.

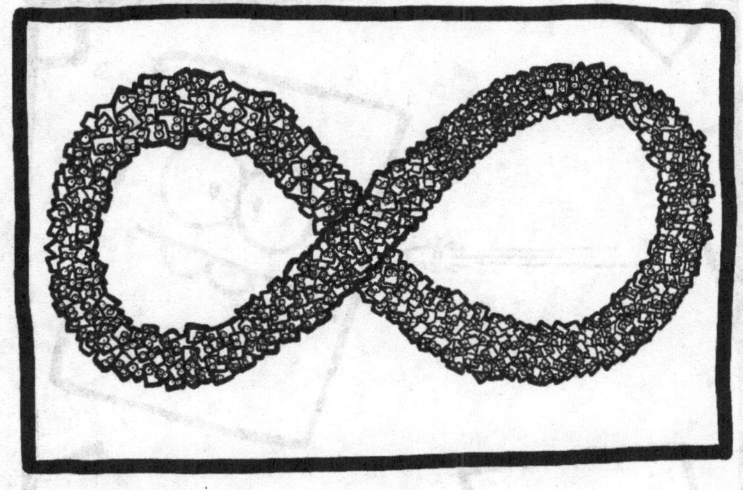

Maybe we'll leave our universe someday...

And go visit another one.

All living things are mostly made of six types of atom. They're called hydrogen, oxygen, carbon, nitrogen, calcium and phosphorus.

Hydrogen atoms came from near the start of the universe. And lots of them sat around in the earliest nebulae.

Oxygen, carbon, nitrogen, calcium and phosphorus came from red supergiants.

Those red supergiants eventually blew up in supernovae...

And shot the atoms across space.

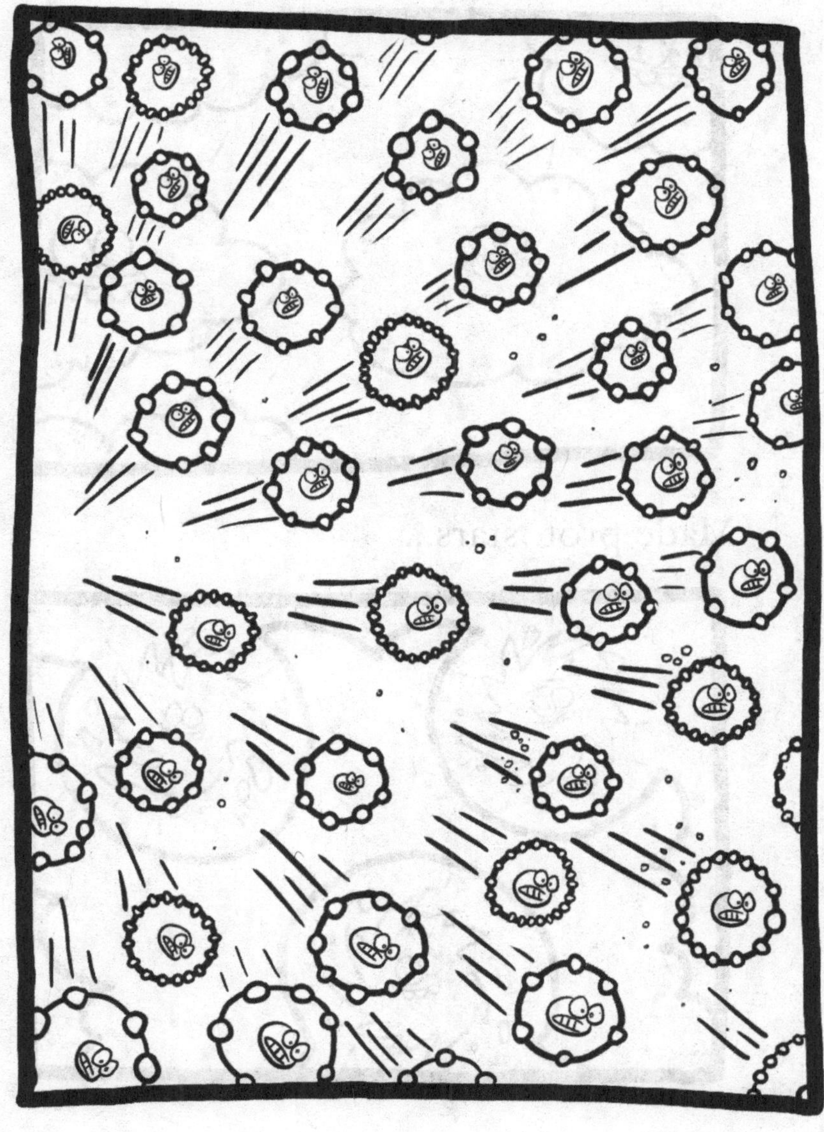

Those atoms formed new nebulae...

Made protostars...

Became stars...

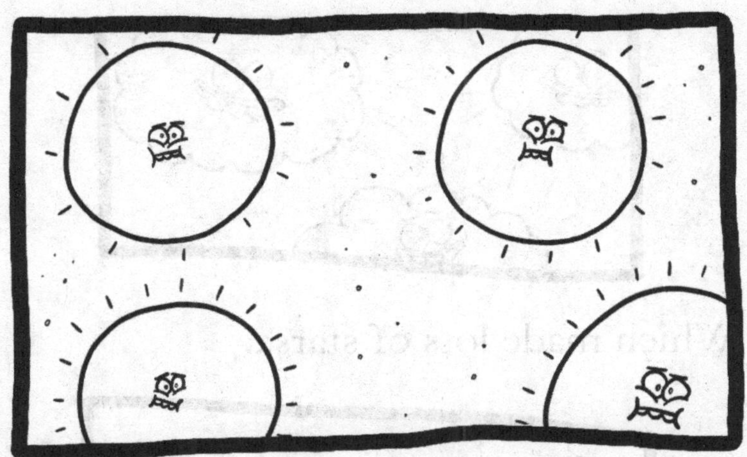

Lots of those stars blew up into supernovae...

Which made lots of nebulae...

Which made lots of stars...

And eventually one of the stars that popped up was the Sun...

Later those atoms made the planets of the Solar System, including Earth.

At first Earth was a hot ball of rock...

But in a while it cooled down and water formed on it.

And then, for the very first time, the atoms made living things...

Living things are all made of little blob shaped things called cells...

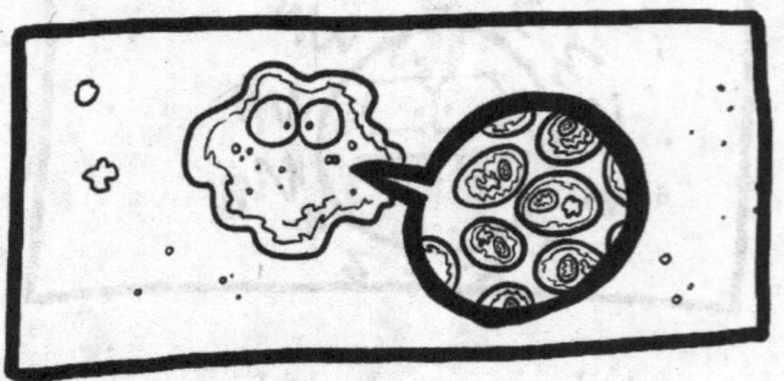

They can grow...

They can react to stuff around them...

Take stuff in and pass it out again...

And they make more living things that are a lot like them...

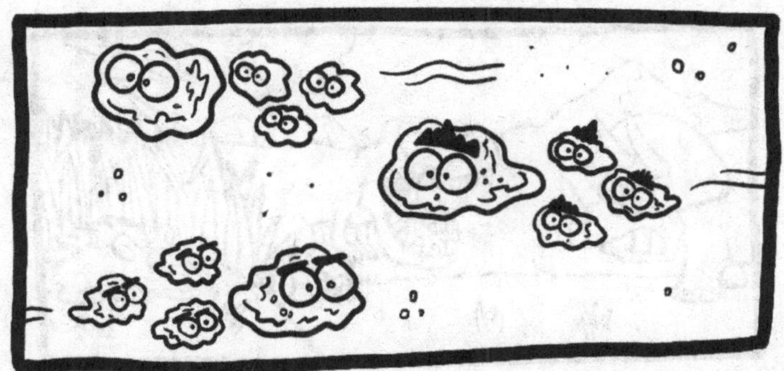

Over time life changed in the water...

Then it went onto the land...

And changed there, too...

Life kept on changing...

Until it changed into people like you.

We have been looking for life outside of Earth but so far we haven't found any...

And that could be because there is no life anywhere other than Earth...

Or there is life somewhere but it can't tell us where it is...

Or there could be life and it could tell us where it is but it doesn't want to...

Or there could be life and it would be very happy to tell us where it is...

But we just haven't found each other yet. It is a big universe, after all.

Stuff In Space

Antimatter: Acts like matter but is not matter.

Asteroid: Really big rock that goes around a Star.

Asteroid Belt: Area full of asteroids between the path of Mars and Jupiter.

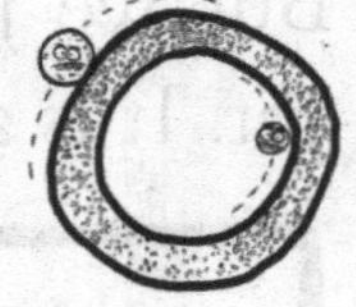

Astronomical Unit: Used to measure distances. Is distance from Earth to the Sun.

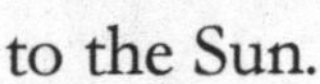

Atom: Small thing made of electrons, protons and neutrons.

Black Dwarf: Forms from a Sun-sized Star that isn't burning or bright anymore.

Stuff In Space

Black Hole:	Giant dark ball in space. Sucks everything in.	 
Brown Dwarf:	Really big and hot planet. Almost a star.	
Calcium:	Atom made in big stars. Lots of it in people.	
Carbon:	Atom made in big stars. Lots of it is in people.	
Cell:	Blobs inside all living things	
Centaur Asteroid:	An asteroid made of ice and rock mixed together.	
Ceres:	The only dwarf planet in the asteroid belt.	

Comet:	Really big chunk of ice that goes through space around a star.	
Core:	Middle of a star where atoms fuse. Planets have cores, too.	
Dark Energy:	Makes gravity that pushes out instead of pulling in. Lots of it.	
Dark Matter:	Invisible, can't touch it. Makes gravity that holds galaxies together.	
Dark Nebula:	Looks black because it blocks light from something else.	
Dead Galaxy	Galaxy. Not making new stars anymore.	

Stuff In Space

Dwarf Planet: Almost a planet. Usually does not have a 'clear path' around a star.

Earth: Rocky planet in the Solar System. You are on it.

Electron: Goes around the nucleus of an atom.

Elliptical Galaxy: Galaxy shaped like a glowing fried egg.

Emission Nebula A nebula that stars grow out of.

Energy: The stuff that lets stuff do stuff like move or get hot.

Eres: Dwarf planet in the Kuiper Belt.

Exo-Planet: A planet that is not in the Solar System.

Galaxy:	Big group of planetary systems, gas and dust.	
Gas:	Matter that is really light and whooshy.	
Gas Giant:	Big planet with most of its mass made of gas.	
Gravity:	Force that pulls stuff toward other stuff.	
Gravity (the other type):	Gravity from dark energy. Pushes away.	
Hadron:	Really, really small thing made from quarks.	
Haumea:	Dwarf planet in Kuiper Belt.	
Heat Death:	End of the universe if everything ran out of light, heat and energy.	

Stuff In Space

Helium:	Atom that is made in stars by fusing hydrogen atoms.	
Hydrogen:	Atom that was made at the start of the universe.	
Ice Giant:	Big planet with most of its mass made of ice.	
Iron	Atom made in big stars. Does not let out energy after fusion.	
Irregular Galaxy:	Galaxy that is not elliptical or spiral shaped.	
Jupiter:	Gas giant planet in Solar System.	
Kilonova:	Explosion when two neutron stars collide.	

Kuiper Belt:	Ring-shaped space full of ice around the edge of the Solar System.	
Lepton:	Really, really small thing. Electrons are a type of lepton.	
Life:	Something that reacts, eats, grows, makes copies of itself and is made of cells.	
Light:	A type of energy that stars shoot out when they are fusing atoms.	
Lightyear	Used to measure distances. How far light travels in a year.	
Liquid:	A type of matter that goes splash and sloshes around.	
Lithium:	A type of atom made in the core of really, really big stars.	

Stuff In Space

Main-Sequence Star:	When a protostar gets round like a ball.	
Makemake:	Dwarf planet in Kuiper Belt.	
Mars:	Rocky planet in Solar System.	
Mass:	Makes stuff heavy or light.	
Matter:	Stuff we see, touch or smell. Solid, liquid, gas or plasma.	
Mercury:	Rocky planet in the Solar System.	
Meteor:	A meteoroid that has entered the air around Earth.	
Meteorite:	A meteor that has hit the surface of a planet.	
Meteoroid:	A small chunk of rock that goes around a star.	

Micrometeor: A micrometeoroid that has entered the air around a planet.

Micrometeorite: A micrometeor that has hit the surface of a planet.

Micrometeoroid: A very small chunk of rock that goes around a star.

Moon: A ball of rock that goes around a planet.

Multiverse Lots and lots of universes all together. Probably exists.

Nebula: Big cloud of gas, dust or plasma in space.

Neptune: Ice giant in Solar System.

Neutron: A type of hadron that is inside the nucleus of an atom.

Neutron Star:	A star core made only of neutrons. Very high mass and gravity.	
Nitrogen:	Atom made in the core of really big stars. Lots of it in people.	
Nucleus:	The middle bit of an atom.	
Oort Cloud:	Ball-shaped cloud of ice that is all around the Solar System.	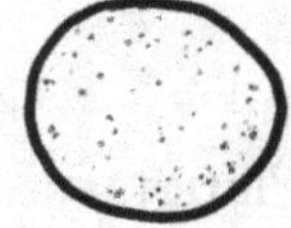
Orbit	The path that something goes on around something.	
Oxygen:	An atom made in really big stars. Lots of it is in the universe.	
Parsec	Used to measure distances. Three and a bit lightyears long.	

Phosphorus	Atom made in big stars. Lots of it is in you.	
Photon:	A very small piece of light.	
Planetary System:	A star and all the things that go round it.	
Planet:	Ball-shaped. Goes on clear path around star.	
Planetary Nebula:	Nebula that forms after a red giant loses its plasma.	
Plasma:	Type of matter which acts like gas but it is very hot.	
Pluto:	Dwarf planet in Kuiper Belt.	
Proton:	Type of hadron inside the nucleus of an atom.	

Stuff In Space

Protostar:	The first part of a Star's life	
Quark:	Really small things that make Hadrons. Go around in threes.	
Red Dwarf:	A really small Star that is hot and bright but not too hot or too bright.	
Red Giant:	Forms from a Sun-Sized Star that has fusing Helium.	
Red Supergiant:	Forms from a really big Star that has run out of hydrogen.	
Reflection Nebula:	A nebula that shines because it is reflecting light from other stuff.	
Remnant Nebula	A Nebula left over after a supernova happens.	

Saturn:	Gas giant in Solar System.	
Scattered Disc	The area between the Kuiper Belt and the Oort Cloud.	
Singularity:	Something with the strongest gravity.	
Solar System	The planetary system we are in.	
Solid:	Matter that keeps its shape.	
Space:	Universe is full of it.	
Spacetime:	Shows how stuff moves in the universe.	
Spiral Galaxy:	Galaxy with arm shapes spinning around.	
Star:	Big, bright balls that fuse atoms in cores.	

Stuff In Space

Sun:	The star in the middle of the Solar System.	
Supernova:	Explosion straight after a big star collapses.	
The Big Bang:	When the scrunched-up universe spread out.	
The Big Crunch:	When the universe squishes in on itself.	
The Big Rip:	When dark energy rips everything apart.	
Universe:	The thing that everything happens in.	
Uranus:	Ice giant in Solar System.	
Venus:	Rocky planet in the Solar System.	
Void:	Spots of the universe that look empty.	

That Book About Space Stuff

Water:	Liquid. Is the thing where life probably started.	
White Dwarf:	Sun-Sized Star that stopped fusing atoms but it's still pretty hot and bright.	
White Hole:	A spot in space that probably shoots out stuff.	
Zombie Galaxy:	A dead galaxy that has taken hydrogen from another galaxy and makes stars again.	

This is Phil. Phil has a PhD in Physics and knows a lot about space. He gave David a lot of help with this book.

So if you read something that is wrong- blame Phil.

Send him all of your complaints. I'll even give you his address and you can mail them to him.

Good luck, Phil.

Also by David Conley...

This...

And this...

THAT BOOK ABOUT GREEK MYTHOLOGY PART 2
WORDS AND PICTURES BY
DAVID CONLEY

Also this...

THAT BOOK ABOUT NORSE MYTHOLOGY PART 1
WORDS AND PICTURES BY
DAVID CONLEY

This, too...

THAT BOOK ABOUT NORSE MYTHOLOGY PART 2
WORDS AND PICTURES BY
DAVID CONLEY

And let's not forget this, either...

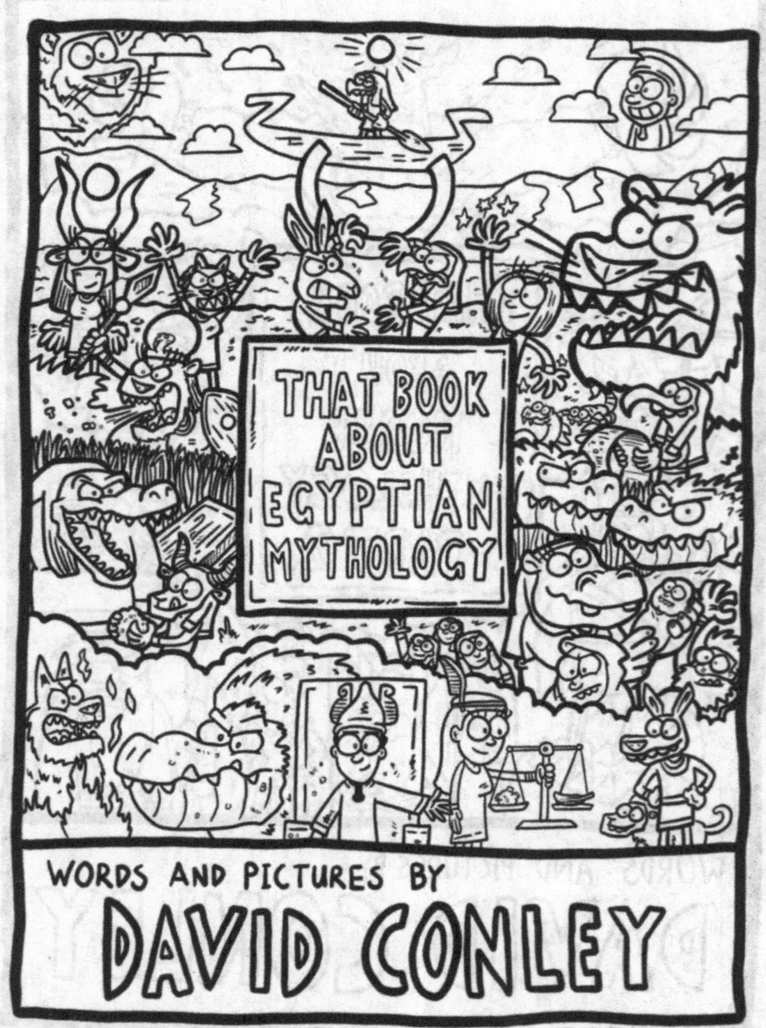

Coming next...

That Book About Life Before Dinosaurs

And after that...

That Book About Australian Parliament Stuff

About The Author

David grew up on the south coast of Australia. Then he kept growing up until he was eventually over 200 thousand kilometres tall. On the way to get milk one Sunday, David tripped over and landed in the Pacific Ocean. He lives out his days as New Zealand and is a very popular tourist destination.

David still loves to draw and write.

Find him on Instagram:

@thatdavidconley

Or just shoot him an email:

thatdavidconley@gmail.com

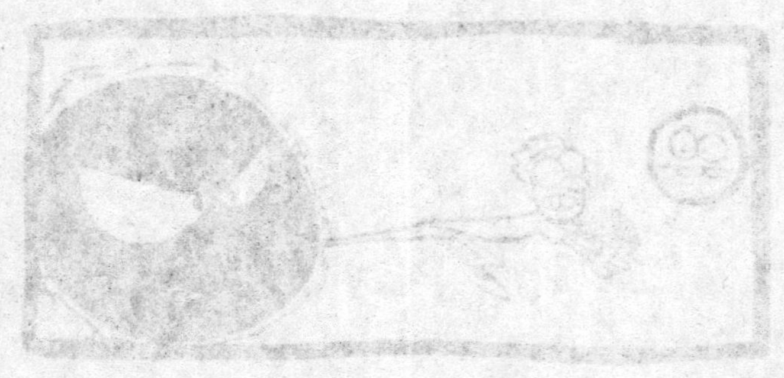

www.ingramcontent.com/pod-product-compliance
Lightning Source LLC
Chambersburg PA
CBHW011158220326
41597CB00026BA/4664

* 9 7 8 0 6 4 5 7 2 4 9 0 5 *